DEUTSCHE FORSCHUNGS- UND VERSUCHSANSTALT
FÜR LUFT- UND RAUMFAHRT E.V. (DFVLR)

SOLAR THERMAL ENERGY UTILIZATION

German Studies on Technology and Application

Editor: M. Becker

Volume 1:
General Investigations on Energy Availability

Springer-Verlag
Berlin Heidelberg GmbH 1987

Dr.-Ing. Manfred Becker
Hauptabteilung Energietechnik der
Deutschen Forschungs- und Versuchsanstalt für Luft- und Raumfahrt e. V. (DFVLR), Köln

ISBN 978-3-540-18028-9   ISBN 978-3-662-01626-8 (eBook)
DOI 10.1007/978-3-662-01626-8

© Springer-Verlag Berlin Heidelberg 1987
Originally published by Springer-Verlag Berlin Heidelberg New York in 1987

2362/3020-543210

Preface

The energy crisis in 1973 and 1979 initiated a great number of activities and programs for low and high temperature application of solar energy. Synthetic fuels and chemicals produced by solar energy is one of them, where temperatures in the range of $600-1000\,^{\circ}C$ or even higher are needed. In principle such high temperatures can be produced in solar towers. For electricity production, the feasibility and operation of solar tower plants has been examined during the SSPS - project (Small Solar Power System) in Almeria, Spain.

The objective of Solar Thermal Energy Utilization is to extend the experience from the former SSPS - program in to the field of solar produced synthetic fuels. New materials and technologies have to be developed in order to research this goal. Metallic components now in use for solar receivers need to be improved with respect to transient operation or possibly replaced by ceramics. High temperature processes, like steam-methane reforming, coal conversion and hydrogen production need to be developed or at least adapted for the unconventional solar operation. Therefore Solar Thermal Energy Utilization is a long term program, which needs time for its development much more time than the intervals expected in between further energy crisis. The "Studies on Technology and Application on Solar Energy Utilization" is a necessary step in the right direction in order to prepare for the energy problems in the future.

Prof. Dr. H. F. Knoche

Rheinisch-Westfälische Technische Hochschule Aachen, Federal Republic of Germany

SOLAR THERMAL ENERGY UTILIZATION

German Studies on Technology and Application

Volume 1: General Investigations on Energy Availability

Contents

YEARLY YIELD OF SOLAR CRS-PROCESS HEAT
AND TEMPERATURE OF REACTION

P. KOEPKE
H. QUENZEL
R. SIZMANN

UNIVERSITÄT MÜNCHEN

Contents   Page

1. Introduction

It is by thermodynamics of advantage to utilize process heat of highest possible and achievable temperature. In conventional thermal electricity power plants the exploited upper temperature is limited by materials properties to about 1000 K. It is believed that e.g., with MHD and thermochemical reactions higher process temperatures can be sustained or handled in new procedures of energy transfer.

In particular in solar energy process heat applications it is the aim to utilize high working temperatures, otherwise a considerable part of the by collection area and devices costly solar energy becomes unavoidably lost as waste heat to remove the process entropy.

However, not only materials to withstand high temperatures pose problems. Also the availibility of highly concentrated solar radiation used to produce high temperatures is terrestrially subject to restrictions and limitations.

Here the solar meteorology and the limited concentrative power of such optically simple systems as heliostat fields with a central receiver cause low solar yields at high concentration levels.

In the present investigation the concerted action of realistic meteorology data, optical heliostat field factor and the chosen temperature level of the process heat is considered (Section 2). In particular the description of the meteorology is shown to be possible and reliable with rather few characteristic parameters (Section 3).

The meteorology is evaluated for Tabernas. The process heat yield and the maximum daily operation hours are then calculated in monthly and yearly averages (Section 4). The obtained data are applied to a specific example of process heat at level 860°C corresponding to the methane steam reforming. The process heat yield is finally presented in Table 5.1 in dependence on the optical concentration factor (Section 5).

2. Evaluation of Available Process Heat

The radiative input through the receiver aperture is the focused solar radiation E_R emerging from the heliostat field area A_F

$$E_R = E \cdot F_F \cdot A_F \cdot F_L$$

E is the time dependent flux density of the incident direct solar radiation; its dependence on meteorology of the atmosphere is discussed and calculated in Secion 3.

The field factor F_F accounts for the efficiency of the heliostat field for focusing the incident solar radiation E into the receiver aperture. F_F is dependent on field design but also on the time varying elevation and azimuth of the Sun. Its structure is discussed in Section 2.3. Here we note that the field factor can be factorized

$$F_F = F_F{}^* \cdot F_{COS}.$$

The factor $F_F{}^*$ is roughly constant in time (which is an approximation only), whereas the cosine factor F_{COS} varies with the position of the Sun.

The landuse factor F_L counts the fraction of the heliostat field land area A_F which at $F_{COS} = 1$ would intercept incident solar radiation. F_L is a measure of the ratio of reflective mirror area to total land area A_F.

The receiver delivers process heat Q_R. The energy balance requires

$$Q_R = E_R - V_R$$

V_R is the total energy loss rate of the receiver: by backscattering of the incident radiation, by thermal emmission and convection through the receiver aperture A_R, by thermal losses through the receiver walls. In the simple case of a perfect black body receiver of internal cavity temperature T_R and with thermal re-emission only

$$V_R = A_R * \sigma T_R{}^4.$$

σ is the Stefan-Boltzmann constant, $5.67 \cdot 10^{-8}$ Wm^{-2}K^{-4}.

We rewrite the balance equation by inserting E_R and separating the geometric and time constant factors

$$Q_R = A_F \cdot F_L \cdot F_F{}^*(E \cdot F_{COS} - V)$$

where

$$V = V_R/(A_F \cdot F_L \cdot F_F{}^*).$$

We finally refer the process heat Q_R to the effective heliostat field area A_R F_L $F_F{}^*$

$$Q = Q_R/(A_F \cdot F_L \cdot F_F{}^*) = E \cdot F_{COS} - V.$$

The units of the varables Q, E, V in this **reduced balance equation** are energy flux densities, Wm^{-2}.

An advantage of the reduced equation is that size and design parameters A_F and F_L and (partially) F_F^* which determine the power level of the solar tower plant are separated from parameters which determine the efficiency of process heat production. Structural design is still present in the cosine factor of the heliostat field, since F_{COS} is contingent upon the pattern of the distribution of the heliostats.

The quantity V, in the following referred to as 'loss coefficient', represents the reduced total loss of the receiver. In particular, V depends on receiver temperature, but likewise we can include in V the parasitic power demand of the solar tower plant. We note that for any particular receiver design
ture T relates to the temperature of the heat extracted from the receiver or consumed in chemical reactions taking place in the receiver

$$V = (V_R(T) + V(parasitic))/(A_F \cdot F_L \cdot F_F^*)$$

In the simple case of a black body receiver with reemission of thermal equilibrium radiation only

$$V = \sigma\, T_R^4/C$$

with the effective geometric concentration factor

$$C = A_F \cdot F_L \cdot F_F^*/A_R.$$

Such a partition of the loss coefficient V, irrespective of its possible complex composition of various loss terms, into two parameters: a ficticious temperature T^* and a geometric concentration ratio C

$$V = \sigma\, T^{*4}/C$$

is rational in rating receiver design and concentrating power of the heliostat-receiver combination.

2.1 Process Heat Yields
We put the yield of process heat in terms of

- the position of the Sun (by day number N of the year and solar time t_S of the day) and of
- the loss coefficient V

$$Q(N, t_S; V)$$

Daily sums are computed

$$Q(N;V) = \int_{t_{on}}^{t_{off}} Q(N,t_s;V)dt_s$$

which then are collected to **monthly averages**

$$Q(V) = \sum_{N=1}^{N_m} Q(N;V)/N_m.$$

(N_m is the number of days of the month in question)

and to **yearly averages**

$$Q(V) = \sum_{N=1}^{365} Q(N;V)/365.$$

The units used are kWh/m^2/day.

The starting time t_{on} and stopping time t_{off} of the daily operation interval are found from

$$Q(V) = 0 = E \cdot F_{COS} - V.$$

t_{on} and t_{off} vary daily because of the changing solar declination
but also because of the seasonally changing meteorological transmittance of the atmosphere. Hence, the monthly (and yearly) average of daily available process heat is subjected twice to the value
of V: in the reduced balance equation $Q = E \cdot F_{COS} - V$, where the
temporary available harvest of Q depends on V, and in t_{on}, t_{of},
where V determines the daily switch-on/off time of the plant.

The monthly and yearly averages of the daily operation intervals
are computed

$$t(V) = \sum_{N=1}^{N_m} (t_{off}(N,V) - t_{on}(N,V))/N_m$$

with N_m the number of days in the month in question, or $N_m = 365$
in case of the yearly average.

The daily available solar energy in the receiver aperture is calculated

$$E(N) = \int_{t_1}^{t_2} E \cdot F_{COS} dt$$

with t_1 and t_2 being times of sunrise and sunset, respectively.

All the averages are calculated separately for clear sky throughout the day (which is ficticious) and for the more realistic day
by including the probability P_C of clouds being present. The daily
meteorological variation of $P_C(N)$ is discussed in Section 3. For
instance

$$Q(V, clear\ day) = \sum_{N=1}^{N_m} Q(N;V)/N_m$$

$$Q(V, cloudy\ day) = \sum_{N=1}^{N_m} Q(N;V)(1 - P_C)/N_m.$$

The same procedure is applied in evaluating the operation time
intervals for clear sky or cloudy sky and for the available solar
energy. No distinction is made between P_C originating from stratiform clouds (i.e., cloud covers of long periods) or convective
clouds (i.e., frequent change in shadowing). No allowance is made
for transient periods of warming up, where process heat of the
required temperature level is not yet available. The consequence
is that the calculated plant operation times (and the process heat
yields) are maximum values. The correction for real operation
times is subject to the thermal behaviour and the chemical engineering of the plant, in particular of the receiver. Once such
explicit corrections are on hand they can be subsequently applied
to the maximum values given here.

Finally, we evaluated the daily solar efficiency of process heat
production

$$\eta\ (N;V) = Q(N;V)/E(N) = Q_R/(A_F \cdot F_L \cdot F_F \cdot E).$$

Note that the efficiency is independent of clouds being absent or
present as long as corrections for transients are excluded.

We computed the monthly and yearly averages of the daily efficiencies according to the procedures defined before.

2.2 Solar Input
The solar input E is calculated according to the principles given
in Section 3. We have used the data collected in Table 2.1.

Examples of numerical values of the various transmittances during
three typical days of a year

Equinox	March	21	solar declination	0	degrees
Summer solstice	June	21	solar declination	23.44	degrees
Winter solstice	December	22	solar declination	-23.44	degrees

are computed in 12 minutes (0.2 hour) intervals for Table 2.2.

Table 2.1 Parameters Used in Irradiance Calculations

(1) Month   (2) Days Per Month   (3) Cloud Probability
(4) Aerosol d_{AE} (5) Water Vapour u_{WA} (6) Ozone u_{O3}
(7) Cirrus Characteristics
    (a)=P_1; (b)=P_2; (c)=P_3; (d)=P_4; (e)=d_1; (f)=d_2; (g)=d_3; (h)=d_4

(1)	(2)	(3)	(4)	(5)	(6)	(7) (a)	(b)	(c)	(d)	(e)	(f)	(g)	(h)
JAN	31	0.35	0.128	1.33	0.28	0.090	0.085	0.125	0.7	0.6	0.06	0.006	0
FEB	29	0.40	0.144	1.33	0.29	0.090	0.085	0.125	0.7	0.6	0.06	0.006	0
MAR	31	0.48	0.160	1.33	0.30	0.050	0.115	0.435	0.4	0.6	0.06	0.006	0
APR	30	0.42	0.220	1.81	0.31	0.050	0.115	0.435	0.4	0.6	0.06	0.006	0
MAY	31	0.29	0.238	1.18	0.30	0.050	0.115	0.435	0.4	0.6	0.06	0.006	0
JUN	30	0.21	0.288	2.29	0.29	0.055	0.020	0.025	0.9	0.6	0.06	0.006	0
JUL	31	0.14	0.259	2.41	0.27	0.055	0.020	0.025	0.9	0.6	0.06	0.006	0
AUG	31	0.18	0.283	2.65	0.26	0.055	0.020	0.025	0.9	0.6	0.06	0.006	0
SEP	30	0.24	0.214	2.41	0.25	0.050	0.030	0.220	0.7	0.6	0.06	0.006	0
OCT	31	0.38	0.147	2.41	0.25	0.050	0.030	0.220	0.7	0.6	0.06	0.006	0
NOV	30	0.40	0.125	1.18	0.25	0.050	0.030	0.220	0.7	0.6	0.06	0.006	0
DEC	31	0.35	0.097	1.45	0.27	0.090	0.085	0.125	0.7	0.6	0.06	0.006	0

Table 2.2 Daily Course of Atmospheric Transmittances in Tabernas

Winter Solstice   December 22

Solar Time hours	t_{RA}	t_{O3}	t_{GA}	t_{WA}	t_{AE}	t_{CI}
0	.8627133	.9759738	.9852708	.8315006	.7407257	.9272998
.2	.8625137	.9759412	.9852634	.8814493	.7403341	.9272168
.4	.8619112	.9758429	.9852413	.8812953	.7391526	.9269668
.6	.8608963	.9756774	.9852041	.8810368	.737162	.9265474
.8	.8594519	.9754416	.9851513	.8806711	.734329	.9259544
1	.8575532	.9751314	.9850825	.8801942	.7306046	.9251819
1.2	.8551671	.9747409	.9849966	.879c011	.7259235	.9242219
1.4	.8522495	.9742624	.9848924	.8788846	.7201986	.9230645
1.6	.848744	.9736859	.9847696	.878036	.7133191	.9216976
1.8	.8445786	.9729987	.9846231	.8770444	.7051434	.9201086
2	.8396623	.9721841	.9844537	.8758959	.6954915	.9182743
2.2	.8338783	.9712204	.9842572	.8745732	.6841336	.9161801
2.4	.8270771	.9700792	.9840298	.8730542	.6707758	.9138006
2.6	.8190653	.9687226	.9837654	.8713109	.6550363	.911108
2.8	.8095896	.9670988	.9834606	.8693067	.6364143	.9080704
3	.7983135	.9651356	.9831031	.8669934	.6142455	.90465
3.2	.7847833	.9627292	.9826819	.8643058	.5876252	.900801
3.4	.7683777	.9597242	.9821792	.8611527	.5553048	.8964644
3.6	.7482345	.9558767	.9815691	.8574022	.5155172	.8915535
3.8	.7231609	.9507787	.9808103	.8528533	.4657065	.8859174
4	.6915932	.9438972	.9798334	.8471784	.4021323	.8792383
4.2	.6520589	.9330753	.9785074	.839791	.3193926	.8707402
4.4	.6081611	.9151574	.9765429	.8294756	.2109893	.8583382
4.6	.628536	.8767496	.9730646	.8129769	7.928972E-02	.8360323

Table 2.2 Daily Course of Atmospheric Transmittances in Tabernas
Continued

Summer Solstice — June 21

Solar Time hours	tRA	tO3	tGA	tWA	tAE	tCI
0	.9173402	.9848136	.9875594	.8981246	.8474631	.9533635
.2	.9172829	.984803	.9875551	.8981003	.8473561	.9533284
.4	.9170852	.984771	.9875455	.8980276	.846974	.9532227
.6	.9167564	.9847175	.9875294	.8979058	.8463334	.9530458
.8	.9162904	.9846418	.9875069	.877344	.8454282	.9527962
1	.9156846	.9845434	.9874778	.8975126	.8442508	.9524722
1.2	.919934	.9844215	.9874413	.8972393	.8427911	.9520719
1.4	.9140309	.9842751	.9873981	.8969134	.8410384	.9515922
1.6	.912969	.9841029	.9873476	.8965331	.8389714	.9510302
1.3	.9117338	.9839036	.9872898	.8956769	.8365782	.9503821
2	.9103291	.9836754	.9872236	.895048	.8338348	.9496434
2.2	.9087271	.9834161	.9871494	.8944301	.8307157	.9483091
2.4	.9069173	.9831234	.9870666	.8937458	.8271905	.9478736
2.6	.9048819	.9827945	.9869745	.8929912	.8232238	.9468298
2.8	.9025997	.9824259	.9868726	.8921621	.8187737	.9456704
3	.9000456	.9820133	.9867601	.8912536	.8137907	.9443388
3.2	.8971962	.9815528	.9866363	.8902596	.8082166	.9429689
3.4	.8939984	.9810379	.9865002	.8891735	.8019822	.9414056
3.6	.8904286	.98462	.9863508	.8879872	.7950651	.9396846
3.8	.8864306	.9799159	.9861867	.8866913	.7871865	.9377914
4	.8819444	.9799924	.9860061	.8852745	.7784075	.9357099
4.2	.8763983	.9782761	.9858074	.8837232	.7685232	.9334223
4.4	.8711958	.9773529	.9855881	.8820212	.7573568	.9309082
4.6	.8647344	.9763029	.9853455	.8801485	.7446886	.9281451
4.8	.8573698	.9753014	.9650759	.8790799	.7302448	.9251077
5	.8489268	.9751014	.984775	.8757842	.713678	.9217683
5.2	.8391789	.973716	.9844371	.8732208	.694625	.9189968
5.4	.8278311	.9721037	.9840548	.8701364	.6722571	.9140601
5.6	.8144926	.9702062	.9836181	.8670599	.6460512	.9096229
5.8	.7986375	.9679418	.9831133	.8632877	.6148827	.9047454
6	.7795458	.9651923	.9825206	.826787	.5773129	.8993796
6.2	.7562216	.961781	.9818104	.8588761	.5313116	.8934524
6.4	.7272895	.9574235	.9809361	.8635984	.4739603	.8868198
6.6	.6910656	.9516429	.9798163	.8476812	.4010366	.8791259
6.8	.6463267	.9433581	.9782947	.8386396	.3067124	.8694017
7	.606948	.931264	.9760074	.826787	.1853343	.8549088
7.2	.7695693	.9697978	.921763	.8671813	4.753065E-02	.8277281

Spring Equinox — March 21

Solar Time hours	tRA	tO3	tGA	tWA	tAE	tCI
0	.9039826	.9326492	.9897341	.9734468	.8214705	.9465716
.2	.9038826	.9826331	.9897296	.9734138	.8212758	.9463208
.4	.9035821	.9825345	.9897162	.8933144	.8206897	.9461681
.6	.9030774	.982503	.9896938	.8931481	.8197055	.9459121
.8	.902363	.9823877	.98962	.8929137	.818312	.9455569
1	.9014308	.9822372	.9848208	.8926098	.8164935	.9450811
1.2	.90027	.9820498	.9847689	.992243	.8142285	.9444958
1.4	.8996666	.9818233	.9847688	.8917848	.8114897	.9437991
1.6	.8972038	.981555	.9866339	.8912579	.8082433	.9429755
1.8	.8952601	.9812414	.9865538	.890699	.8044471	.9420208
2	.8930097	.9808784	.9864587	.8899562	.8000591	.94923
2.2	.8904208	.9804608	.9863506	.8891712	.79499	.9398908
2.4	.8875552	.9799822	.9862284	.8882983	.7891906	.938273
2.6	.8840659	.9794351	.984091	.8872997	.7825597	.9366886
2.8	.8801955	.9788698	.9859568	.8861957	.7749837	.9349109
3	.8757729	.9780943	.9857638	.9849648	.7683228	.9329211
3.2	.8701102	.9772739	.9855976	.8835931	.7544032	.9306989
3.4	.8640988	.9762292	.9853514	.8820633	.7459071	.9282134
3.6	.8581919	.9752358	.9851056	.8803542	.7318578	.9254409
3.8	.8504154	.973761	.9848274	.8784389	.7165996	.9223462
4	.8413314	.9724611	.984109	.8762832	.6987683	.9182998
4.2	.8306261	.9706759	.9841148	.8739421	.6777467	.9156313
4.4	.8178753	.9685199	.9837278	.8710559	.6528983	.910718
4.6	.8024918	.9658672	.9832348	.8678422	.6224616	.9058938
4.8	.7836441	.9625238	.9823467	.8644834	.5853823	.8994297
5	.760132	.9581722	.9817288	.8598037	.5390345	.8944044
5.2	.7392105	.9522521	.9810242	.8541235	.4797558	.8874501
5.4	.691482	.9436604	.9798298	.8471583	.4019053	.8792151
5.6	.6420128	.9298457	.9781306	.8377556	.2970654	.8693765
5.8	.5986263	.9030294	.9753592	.8236905	.1568688	.8507286
6	1.338819	.8187481	.9689838	.795817	1.50525E-02	.81157

The solar declinations ϑ (in daily intervals) were obtained from The Astronomical Almanac for the year 1977. The latitude of Tabernas is taken ϕ = 37.1 degrees; the altitude of the Plataforma Solar de Almeria is assumed to be such as to correspond with an average atmospheric pressure of p = 950 hPa.

The solar elevation h is calculated by

$$\sin h = \cos\vartheta \cdot \cos\phi \cdot \cos\omega + \sin\vartheta \cdot \sin\phi.$$

The solar hour angle ω is

$$\omega = (t_S/4) \quad \text{(degrees)}$$

with t_S = 0 at solar noon; ω is daily adjusted to 0 at solar noon.

The solar azimuth is calculated from

$$\sin\jmath = -\sin\omega \cdot \cos\vartheta / \cos h.$$

The standard relative air mass AMS is calculated from Kasten's formula, Eq.(3.6).

The daily variation of the solar constant E_{SC} = 1367 W/m^2 is allowed for with

$$E_0 = 1367(1 + 0.034 \cdot \sin(2\pi(82.8 - N)/365.25)) \quad (W/m^2)$$

where N is the day number of the year, N = 1 on January 1st.

2.3 Field Factor

The geometric optical field factor F_F (field efficiency) is a product of several separate efficiencies (Stine - Harrigan 1985)

$$F_F = F_{COS} \cdot F_{REF} \cdot F_{ATM} \cdot F_{SHA} \cdot F_{BLO} \cdot F_{CAP}.$$

The indices have the meaning
COS: cosine factor (see Section 2.4)
REF: reflectance of heliostat mirrors
ATM: optical transmission of radiation between heliostats and receiver aperture; it depends on the atmospheric transmittance at ground level
SHA: fraction of solar radiation incident on a heliostat escaping shadowing by other heliostats
BLO: fraction of reflected radiation escaping being blocked by other heliostats on its path between heliostat and receiver aperture
CAP: fraction of concentrated radiation captured through the receiver aperture.

We have neglected other factors contributing to lessen F_F, e.g., shadowing of the heliostats by the tower.

The various F_i depend on
- incidence direction of the solar beam
- materials properties (e.g., the reflectance)
- distance between heliostats and tower
- field design

In general most of the F_i are dependent on incidence angles (declination and azimuth) and quality (natural divergence, aureole) of the beam.

In particular the cosine factor F_{COS} changes with incidence angle, whereas the capture fraction is less variable over wide azimuth intervals in many heliostat field - receiver aperture designs. We therefore, as an approximation, collect five of the six F_i in a single, constant F_F^*, leaving separate the cosine factor

$$F_F = F_{COS} \cdot F_F^*.$$

F_F^* is of order of 1; in a well designed field it will be about $(0.95)^5 = 0.77$.

The land use factor F_L of the SSPS - CRS - installation is close to 0.25. The small value guarantees low shadowing and blocking factors in the field design.

2.4 The Cosine Factor
We calculate the cosine factor by

$$F_{COS} = \cos \Theta_i.$$

Θ_i is the incident angle of the solar beam on a heliostat mirror which is correctly oriented to reflect the beam into the receiver aperture.

We use the geometric relation (Stine - Harrigan 1985)

$$\cos 2\Theta_i = (H \cdot \sin h + e \cdot \cos h \cdot \sin \gamma + n \cdot \cos h \cdot \cos \gamma)/(H^2 + e^2 + n^2)^{0.5}$$

H is the height of the receiver aperture on top of the tower; e and n are east and north orthogonal coordinates of the heliostat positions in the horizontal heliostat field area A_F.

Any particular geometry of heliostat field layout can be calculated. For simplicity we have used a circular field area of radius R_F with regularly distributed heliostats, see Figure. 2.1.

It suffices to divide the area into segments (in the present calculation 60 were used) and to assign typical heliostat locations to each segment (see for instance the circled position in the hatched segment area). Any number of heliostats in such a segment is then represented by a typical heliostat position.

We compute the area weighted average of the cosine factor of the 60 positions for any solar declination h and azimuth γ (i.e., for any time t_s, for every day number N of the year).

Fig.2.1 Circular distribution
of heliostats.   Center: solar
tower.   The circles represent
heliostats with locations typi
cal for a   segment   (e.g., the
hatched area).   In total 60
segments are used in the calculation of the cosine factor.

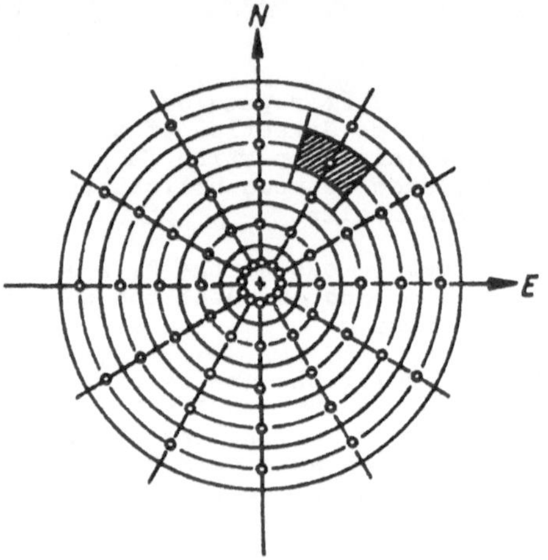

We have assumed the tower height H to be equal to the field radius
R_F. Then, at the rim of the heliostat field the angle of beam elevation from the heliostat towards the receiver aperture is 45 degrees.   Such a field design corresponds to maximum power (see Section 2.5).

The cosine factor F_{COS} which in general for a circular field
contains   as   characteristic parameter the ratio H/R_F,   is in   the
special case because of $H/R_F = 1$ not anymore dependent on geometric dimensions.

For   illustration   the   daily course of the   solar   radiation E_{he}
emerging from the heliostats, being focused into the receiver
aperture is plotted in Figure 2.2,   together with the cosine factor. The plot is valid for Tabernas at spring equinox. Fig. 2.3 is
the   course for the heliostat field factor (at 100%   reflectivity)
as reported from HELIOS measurements (Schiel, 1983). The agreement
between   the calculated and the measured cosine factor   is better
than   could   be expected regarding the different field designs in
question.

2.5 Concentration Factor
The reduced   receiver   loss coefficient V includes a geometric
concentration factor $C = A_F \cdot F_L \cdot F_F^* / A_R$.   The two areas A_F and A_R
are related to each other for given receiver aperture size. In the
simple case of a down facing cavity receiver (see Figure 2.4)   and
the condition of A_R to accept all the radiation within a divergence angle of ß (half angle) emerging from the outmost heliostats
it follows

$$A_R = \pi R_F^2 \tan^2 \beta / \cos^2 \psi \, \sin^2 \psi \, .$$

Here πR_F^2 is the area A_F of the circular field with evenly   distributed heliostats.

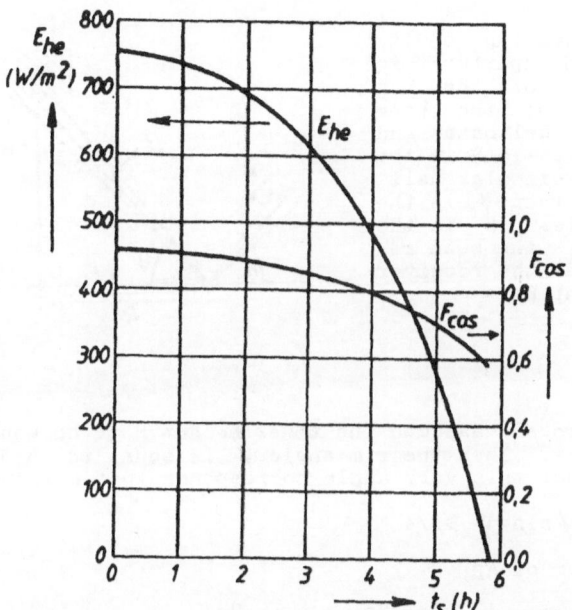

Fig.2.2 Computed concentrated solar radiation flux density E_{he}
emerging from the circular heliostat field versus lapse of solar
day time (left hand ordinate scale). The cosine factor of that
particular heliostat distribution is also plotted (right hand
ordinate scale). The data are valid for March 21, Tabernas Plataforma Solar.

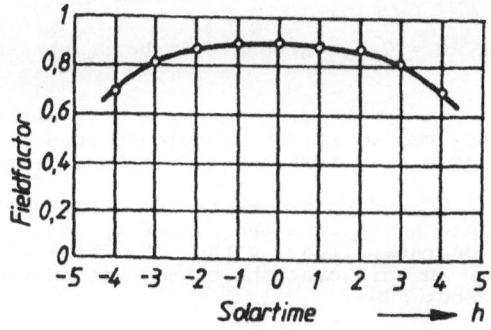

Fig.2.3 Field factor measured for the SSPS Tabernas installation
with HELIOS (Schiel 1983), adjusted to 100 % reflectivity of the
heliostat mirrors. The data are valid for beginning of October

Fig.2.4 Downfacing cavity receiver with aperture A_R. The elevation of the Sun is the angle h; the beam angle between heliostat and receiver is ψ. If R is the radius of the circular heliostat field, then ψ is the field rim angle. ß is the divergence of the beam reflected towards the receiver at tower height H

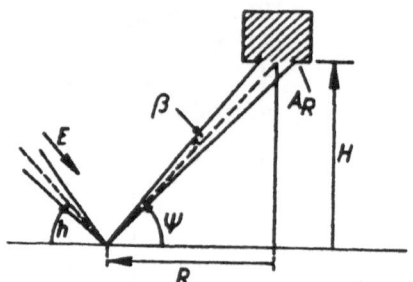

We have previously assumed the tower height H to be equal to the field radius R_F. Then the rim angle ψ is equal to 45 degrees. It can be seen that this very angle corresponds to a maximum of

$$\cos^2 \psi \, \sin^2 \psi \rightarrow 1/4$$

or to a minimum of A_R

$$A_R(min) = 4\pi R^2 \tan^2 ß = 4A_F \tan^2 ß$$

Hence, the geometric concentration factor for solar radiation becomes

$$C = F_L \cdot F_F^* / (4 \cdot \tan^2 ß).$$

As a numerical illustration, we take

$$ß = 0.6 \text{ degrees} \qquad F_L = 0.25 \qquad F_F^* = 0.8.$$

Then   C = 460.
The reason for using ß = 0.6 degrees is it being the sum of three contributions

$ß_1$ = 0.27 degrees: astronomic beam divergence
$ß_2$ = 0.2 degrees: mirror misorientations
$ß_3$ = 0.15 degrees: turbulence in ground level air.

The numbers given for $ß_2$ and $ß_3$ are estimates. For small field areas and well designed heliostats the sum of $ß_2$ and $ß_3$ can be reduced to about 0.15 degrees, leaving the total beam divergence at about 0.45 degrees. In this case the concentration factor is enhanced from 460 to about 800.

With other receiver geometries similar concentration factors are possible. In particular for wide fields (high power level receivers) the open, external receiver has the advantage of not being limited to the 45 degrees power optimum, see Fig.2.5.

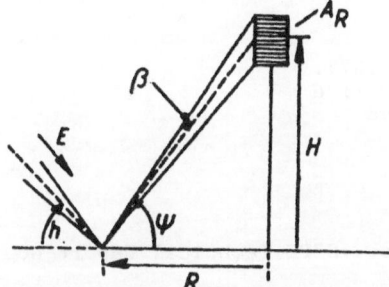

Fig.2.5 External receiver surface
The symbols correspond to the
details given with Fig.2.4

A terminal concentrator can double the concentration factor, lea-ving us with possible C = 900 up to 1600. Fig.2.6 illustrates the terminal concentrator proposals for the two receiver designs.

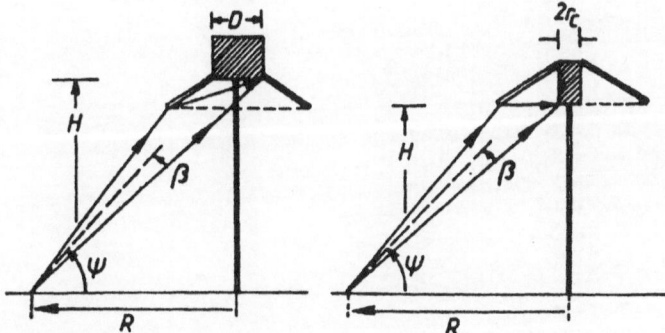

Fig.2.6 Terminal concentrator proposals with enhancement in con-centration by about 2

References

The Astronomical Almanac for the Year 1977.
Washington: U.S. Government Printing Office
London: Her Majesty's Stationery Office

Stine, W.B. and R.W. Harrigan (1985)
Solar Energy Fundamentals and Design.
John Wiley and Sons, New York

Schiel, W. (1983) HERMES Measurements. SSPS Technical Report No. 4/83, Paper 2.6

Fricker, H.W. (1983). A proposal for a novel type of solar gas receiver. In: International Seminar on Solar Thermal Heat Production and Solar Fuels and Chemicals. DFVLR - Stuttgart, October 13 - 14, 1983. Paper No. 11

Müller, W.D. (1986) LURGI, Verfahrenstechnik · Ingenieurtechnik · Anlagenbau, Frankfurt, D-6000. Private Communication

3. Incident Solar Radiation

3.1 General

The radiant flux density from the Sun $E(t,\Theta)$ which is to be re-flected into the receiver is investigated to its dependence on time t and angle Θ of the receiver field of view. The radiation can be split into two components, E_{sol} and E_{aur}. Here, E_{sol} is the attenuated direct solar radiation plus the radiation scattered into the solid angle of the solar disc $0 \leqslant \Theta \leqslant 0.267^\circ$ (see Section 3.3); $E_{aur}(\Theta)$ is the circumsolar radiation (aureole) consisting of scattered radiation in the ring region beyond 0.267° up to Θ. Both E_{sol} and E_{aur} are measured perpendicularly to the direction of propagation of E_{sol}

$$E(t,\Theta) = E_{sol}(t) + E_{aur}(t,\Theta) \tag{3.1}$$

The time dependency of E arises from the change in the Sun's ele-vation in the course of the day and the year, i.e., from the path length of the radiation through the atmosphere. It also arises from the variations in the many meteorological parameters that cause attenuation and scattering (clouds, aerosol particles, water vapour, ozone and other gases). The time variation of these quan-tities will be treated in Sections 3.5 and 3.6, whereas Sections 3.2, 3.3 and 3.4 present relationships between E and the meteoro-logical parameters in analytical forms.

3.2 Attenuated Direct Solar Radiation

The attenuation of the solar radiation on its path through the atmosphere to the ground level is described in terms of **transmit-tances** t_i arising from different extinction processes

$$E_{sol} = E_{sc} \cdot f_1 \cdot f_2 \cdot f_3 \cdot t_{RA} \cdot t_{O3} \cdot t_{GA} \cdot t_{WA} \cdot t_{AE} \cdot t_{CI} \tag{3.2}$$

The indices have the following meaning:

RA   Rayleigh scattering by the molecules of the air
O3   absorption by ozone
GA   absorption by atmospheric gases of constant mixing ratio
WA   absorption by water vapour
AE   extinction by aerosol particles
CI   extinction by cirrus clouds.

Since thick clouds exhibit a transmittance t_C that is effectively zero, their influence will be treated separately in this Section.

Eq.3.2 presents the attenuated direct solar radiation available on the heliostat mirrors for being focused into the receiver. If the receiver has an aperture larger than the image of the Sun then the aureole radiation becomes of importance, see Section 3.3.

Although the optical parameters of scattering and absorption are wavelength dependent, they will be treated here as empirically ad-justed and spectrally averaged values. Therefore, the presented analytical expressions are approximate formulae only

The Solar Constant (the extraterrestrial solar radiant flux density at the mean Sun-Earth distance) has the value (Fröhlich, 1985)

$$E_{sc} = 1367 \ Wm^{-2}.$$

The approximate transmittance functions by Bird and Hulstrom (1981) correspond closely to transmittances which were calculated from theoretical relationships. In the following we will use these values and quote them as B+H. The original publication by Bird and Hulstrom employs a Solar Constant of $E_0 = 1353 \ Wm^{-2}$. For $E_0 = 1367 \ Wm^{-2}$ a correction of the B+H data must be made. According to Iqbal (1983) the appropriate correction factor f_1 in Eq.3.2 is

$$f_1 = 0.9751.$$

Apart from this adjustment, the spectrally weighted transmittance functions depend on the spectral distribution of the incoming solar radiation. A comparison by Iqbal (1983, Fig.7.76) shows that the radiant flux densities would be 2% higher if calculated with the new solar spectrum of WRC (World Radiation Center), rather than with the NASA spectrum employed by B+H. This produces a correction factor

$$f_2 = 1.02.$$

The Earth-Sun distance varies by about 3% during the year which in Eq.3.2 is accounted for by (see Section 2.2)

$$f_3 = (\overline{D}_{ES}/D_{ES})^2 \tag{3.3}$$

with $\overline{D}_{ES}$ the mean distance.

The transmittance related to Rayleigh scattering at air molecules is according to B+H

$$t_{RA} = \exp(-0.093 \cdot (1 + AM - AM^{1.01}) \cdot AM^{0.84})) \tag{3.4}$$

with AM the relative air mass, weighted by the actual pressure p

$$AM = AMS \cdot (p/p_0) \tag{3.5}$$

($p_0 = 1013$ hPa). According to Kasten (1966) the standard relative air mass AMS at solar elevation h is

$$AMS = (\sin h + 0.15 \cdot (3.885 + h)^{-1.253})^{-1} \tag{3.6}$$

Table 3.1 lists AMS for $p_0 = 1013$ hPa and AM for $p = 950$ hPa, the local mean air pressure in Tabernas at 500 m above sea level.

Table 3.1 Standard relative air mass AMS and atmospheric pressure weighted relative air mass AM = AMS·(p/p₀) for p₀ = 1013.2 hPa and p = 950 hPa ·(corresponding to the altitude of Tabernas, 500 m) for several solar elevation angles h and zenith distances z = 90° - h

h	z	AMS	AM
0°	90°	36.51	34.23
5°	85°	10.32	9.68
10°	80°	5.58	5.23
20°	70°	2.90	2.72
30°	60°	1.99	1.87
40°	50°	1.55	1.46
50°	40°	1.30	1.22
60°	30°	1.15	1.08
70°	20°	1.06	1.00
80°	10°	1.01	0.95
90°	0°	1.00	0.94

The absorption by ozone is in B+H

$$t_{O3} = 1 - 0.1611 \cdot U_{O3} \cdot (1 + 139.48 \cdot U_{O3})^{-0.3035} \qquad (3.7)$$

Here we have omitted a small correction term which is present in the original B+H (1981) expression.

$$U_{O3} = u_{O3} \cdot AMS \qquad (3.8)$$

u_{O3} is the vertical thickness of the ozone layer in cm(STP).

The transmittance of the gases of constant mixing ratio, CO_2, N_2 and O_2 is after B+H

$$t_{GA} = \exp(0.0127 \cdot AM^{0.26}) \qquad (3.9)$$

The transmittance od water vapour is according B+H (1981)

$$t_{WA} = 1 - 2.4959 \cdot U_{WA}/((1 + 79.034 \cdot U_{WA})^{0.6828} + 6.385 \cdot U_{WA}) \qquad (3.10)$$

with

$$U_{WA} = u_{WA} \cdot AMS. \qquad (3.11)$$

U_{WA} is the water vapour content in the vertical column above the site in $g \cdot cm^{-2}$ and AMS denotes the standard relative air mass.

The transmittance as affected by aerosol particles can be measured by their spectral optical thickness $d_{AE}(\lambda)$, by vertical turbidity or by horizontal visibility. The uncertainty in the aerosol transmittance calculated from these different input data increases in the given sequence. However, the visibility range VIS (in km) is the most often available measure and it suffices for rough estimates (Iqbal, 1983)

$$t_{AE} = (0.97 - 1.265 \cdot (VIS)^{-0.66}) \cdot AM^{0.9} \qquad (3.12)$$

A better measure is the spectrally integrated turbidity. Since the spectral optical properties of aerosols vary according to type, it is appropriate to measure them at more than one wavelength. This approach is applied in Eq.3.13, where the integrated d_{AE}-values are composed of spectral optical thicknesses at wavelengths 0.38 and 0.5 μm. In this approach, the aerosol in the vertical column above the site is rated. Hence, standard relative air mass AMS enters the equation, not the pressure weighted relative air mass

$$t_{AE} = \exp(-d_{AE}^{0.873} \cdot (1 + d_{AE} - d_{AE}^{0.7088}) \cdot AMS^{0.9108}) \quad (3.13)$$

with

$$d_{AE} = 0.2758 \cdot d_{AE}(\lambda = 0.38 \text{ μm}) + 0.35 \cdot d_{AE}(\lambda = 0.5 \text{ μm}) \quad (3.14a)$$

or, if data are only available at 0.5 μm

$$d_{AE} = 0.7 \cdot d_{AE}(\lambda = 0.5 \text{ μm}) \quad (3.14b)$$

A further transmittance function t_{CI} is necessary to account for the attenuation of the solar radiation by cirrus clouds. We divide them into three classes. Only one of the cirrus classes is assumed to be present, since usually only one cirrus layer exists, if at all. Then the time-averaged transmittance t_{CI} depends on the individual transmittances $t_{CI}^{(i)}$ of cirrus clouds and their probabilities $p_{CI}^{(i)}$ in the respective classes 1, 2 or 3. With a fourth class of completely clear sky, $t_{CI}^{(4)} = 1$, the average transmittance is

$$t_{CI} = \sum_{i=1}^{4} t_{CI}^{(i)} \cdot p_{CI}^{(i)} \quad (3.15)$$

with $\sum_{i=1}^{4} p_{CI}^{(i)} = 1$ and $t_{CI}^{(4)} = 1$ for the case without cirrus.

The transmittance of cirrus is taken as being independent of the wavelength, which is a reasonable approximation. Thus,

$$t_{CI}^{(i)} = \exp(-d_{CI}^{(i)} \cdot AMS) \quad (3.16)$$

with $d_{CI}^{(i)}$ the optical thickness and AMS the standard relative air mass. Eq.3.16 does not contain the fact that for cirrus clouds the attenuation of the radiation by extinction is partly compensated by radiation that is forward scattered into the angular region of the solar disc. The disc radiation is increased in the order of 10 % up to 20 % and more as the optical thickness and relative air mass increase. These figures are derived from Thomalla et al. (1983). We can correct Eq.3.16 for the scattered radiation. However, the uncertainty in the actual optical thickness of cirrus is greater than the correction would bring about in Eq.3.16. Aerosol particles scatter radiation in forward direction, but less than cirrus; we shall disregard this effect in the following.

Thick clouds - which will here be referred to as "clouds" - attenuate the direct solar radiation so strongly that a focusing solar power device is put off operation.

This is the case when the optical thickness d_C of the cloud is greater than 2. Accordingly, the transmittance of all such clouds is set zero. Two terms are left.

$$t_C(1) = 0 \quad \text{if clouds are present}$$
$$t_C(2) = 1 \quad \text{if clouds are absent.}$$

It follows for the cloud transmittance

$$t_C = 1 - p_C.$$

p_C describes the probability of there being thick clouds between the Sun and the heliostat field. t_C is used in Section 3.7 to calculate time-averaged radiant flux densities. However, in solar thermal the very number of daily hours is required, where the flux density at the receiver is greater than a given value. In the following we calculate first by applying Eq.3.2 the daily receiver operation hours without clouds and then multiply by $(1 - p_C)$ to obtain the actually available hours. This implies that the real day time of appearence of clouds does not matter.

Fig.3.1 (Fig.7.6.1 of Iqbal, 1983) shows the atmospheric transmittance as a function of the air mass for an atmosphere of high turbidity (VIS = 10 km), without cirrus. Above a solar elevation of 30° the effect of gaseous absorbers is small (the transmittance is about 0.8). The major attenuation is due to aerosol particles and also to cirrus. For solar elevations below 30° the transmittances decrease rapidly because of the steep increase of the air mass.

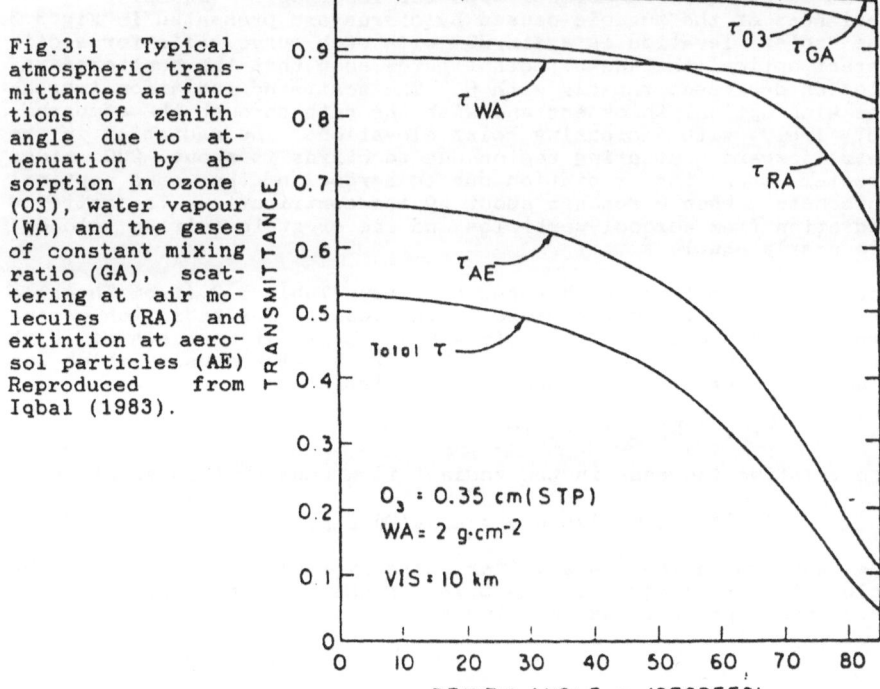

Fig.3.1 Typical atmospheric transmittances as functions of zenith angle due to attenuation by absorption in ozone (O3), water vapour (WA) and the gases of constant mixing ratio (GA), scattering at air molecules (RA) and extintion at aerosol particles (AE) Reproduced from Iqbal (1983).

3.3 Energy Gain by Increasing the Receiver Field of View

Part of the radiation which is removed from the direct solar beam by extinction is forward scattered towards the Earth's surface. Aerosol particles and in particular the ice crystals in cirrus clouds are the species which scatter preferentially in forward direction. Hence, scattered radiation is especially strong at and near the Sun's disc. This causes a brightening of the sky around the Sun which is called the **aureole**. Its inner edge is bounded by the disc of the Sun. Moving radially outward, the aureole fades gradually until it merges with the brightness of the sky.

Increasing the angle Θ of the field of view of the receiver aperture increases the usable incident radiation. The radiant flux density of the aureole $E_{aur}(\Theta)$ has been calculated assuming single scattering of the radiation by Thomalla et al. (1983) for various aerosol types as well as cirrus of different thicknesses and including the absorption in the atmosphere. The assumption of single scattering is particularly justified in the near forward scattering region.

$$E_{aur}(\Theta) = 2\pi \int_{\Theta_R}^{\Theta} L(\Theta)\sin\Theta d\Theta \qquad (3.18)$$

with $L(\Theta)$ the radiance at an angular distance Θ from the center of the Sun due to scattered radiation, accounting for the Sun not being a point source, but a disc of angular radius $\Theta_R = 0.267°$. Fig.3.2 shows $L(\Theta)$ for solar elevation 20° and three values of the aerosol thickness d_{AE}. "Continental background aerosol" was chosen as it is a realistic aerosol type for Tabernas. Examples of the radiances of the aureole caused by cirrus are presented in Fig.3.3. The solar elevation is again 20° with each curve valid for a different optical thickness. Both figures show that the scattered radiation decreases rapidly with Θ. The scattered radiation increases with optical thickness and with the path through the atmosphere, i.e., with decreasing solar elevation. The radiation in the near forward scattering region due to cirrus is about 200 times greater than the radiation due to aerosol of the same optical thickness. When Θ reaches about 5° the contributions to scattered radiation from aerosol particles and ice crystals (cirrus clouds) are nearly equal.

For several optical thicknesses of cirrus Table 3.2 (from Thomalla et al., 1983) lists the radiant flux density for radiation coming from the Sun disc region, $E(\Theta = \Theta_R) = E_{sol}$ (it consists of both direct and scatterd radiation) and for the flux density which includes the aureole up to the angular distance Θ

$$E(\Theta) = E_{sol} + E_{aur}(\Theta).$$

The relative increase in the radiant flux density is also given

$$(E(\Theta) - E(\Theta_R))/E(\Theta_R) = E_{aur}(\Theta)/E_{sol} \qquad (3.19)$$

The numbers in Table 3.2 differ slightly from those which can be calculated from Eqs.3.13 and 3.16, since these are derived from integrated spectral radiant flux densities.

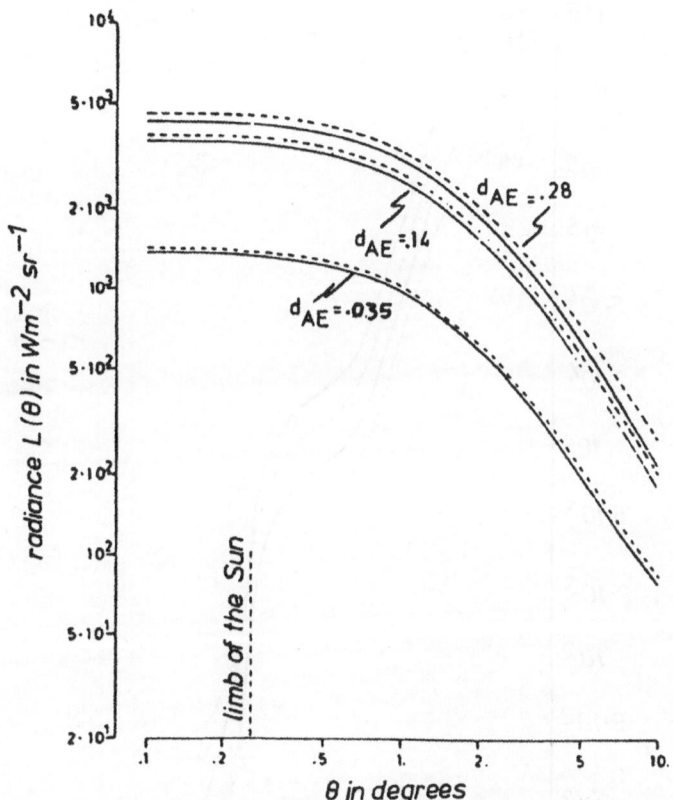

Fig.3.2.   Radiance L(Θ) due to scattered solar radiation as function of aureole angle Θ at solar elevation 20° for continental
background aerosol with optical thickness 3.   The solid curves account for single scattering and the dashed curves for multiple
scattering. After Thomalla et al. (1983)

Atmospheres containing aerosol particles but no cirrus produce an
aureole up to 5° which is enhanced in radiant flux density only if
the Sun is low i.e., the transmission along the optical path is
small.   But then the absolute values of E(Θ) are so small that it
is hardly productive to run a solar device.   It is, therefore, not
worthwhile to consider an increase of the receiver aperture solely
in view of aerosol forward scattering.

However, there is a remarkable increase in the radiant flux density when the cirrus induced aureole is included.   The scattering
function has a steep slope; the increase is given in Table 3.2 for
several aureole angles Θ.   For Θ = 5° radiant flux densities are
greater than 300 Wcm^{-2}, which is an enhancement of more than 30 %.

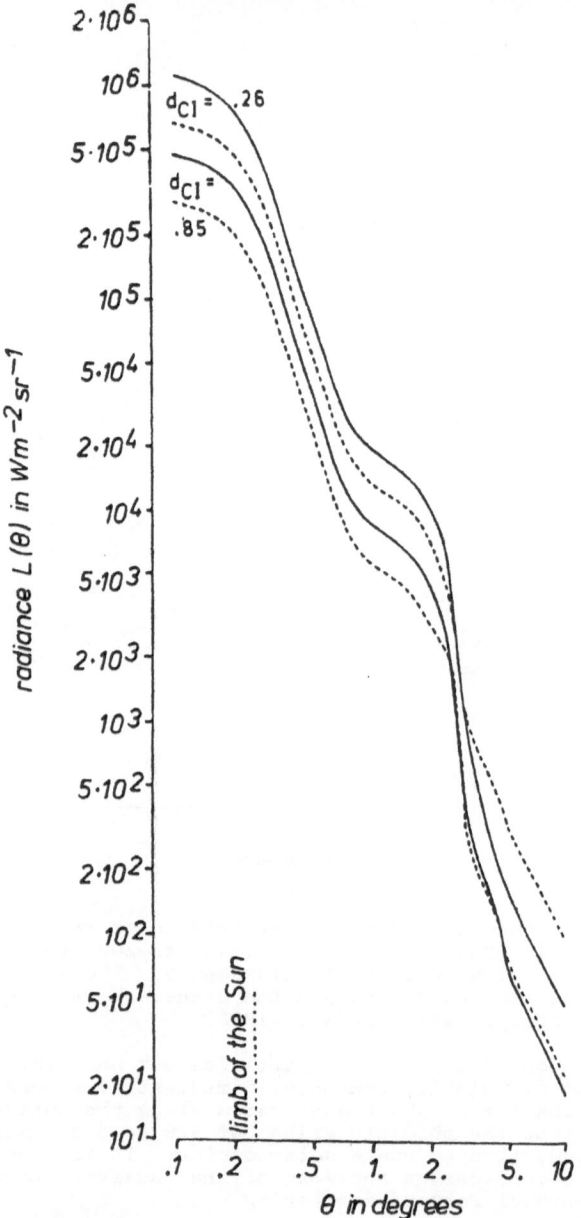

Fig.3.3   Radiance $L(\theta)$ due to single scattered solar radiation as function of aureole angle θ at solar elevation 20° for cirrus clouds of two optical thicknesses 0.26 and 0.85 (solid lines) and with additional aerosol particles of optical thickness 0.14 (dashed lines). After Thomalla et al. (1983)

Table 3.2   Incident radiant flux density by direct solar radiation $E(\Theta = \Theta_R = 0.27°) = E_{sol}$ and by direct and scattered solar radiation $E_{sol} + E_{aur}(\Theta)$ for several aureole angles Θ, depending on the solar elevation angles h, the aerosol optical thickness d_{AE} and the optical thickness of cirrus clouds d_{CI}. The relative increase of the incident radiant flux density due to the aureole is also given. After Thomalla et al. (1983)

h	d_{CI}	d_{AE}	$E(\Theta_R)$ $=E_{sol}$	(1)=E(Θ) $=E_{sol}+E_{aur}(\Theta)$		(2)=(E(Θ)-E(Θ_R))/E(Θ_R) in %					
				$\Theta=$ 0.5°		1°		5°		10°	
			Wm^{-2}	(1)	(2)	(1)	(2)	(1)	(2)	(1)	(2)
60°	0.26	0.0	774	816	5.4	836	8.0	897	15.9	902	16.5
	0.43		640	796	8.7	722	12.8	803	25.5	809	25.4
	0.85		395	458	15.9	487	23.3	578	46.3	584	47.9
	0.26	0.14	641	676	5.5	694	8.3	757	18.1	770	20.2
	0.43		530	576	8.8	599	13.1	676	27.6	688	30.0
	0.85		326	379	16.0	404	23.7	485	48.5	494	51.4
20°	0.26	0.0	396	445	12.8	469	18.7	541	37.1	547	38.5
	0.43		241	288	19.5	308	27.9	374	55.4	379	57.5
	0.85		68	90	32.2	100	47.1	131	93.0	133	96.2
	0.26	0.14	248	280	12.9	296	19.4	352	42.3	364	47.0
	0.43		151	180	19.6	195	29.2	243	61.4	251	66.5
	0.85		42	56	32.5	63	48.2	84	98.3	86	104.1

Fig.3.4 shows the radiant flux density E_{aur} contained in the aureole extending to $\Theta = 5°$ as a fraction of E_{sol}. It measures the relative increase in incident radiation with increasing receiver aperture. To estimate the absolute change in the radiation as d_{CI} increases (and for comparison with Fig.3.9), we include a demarcation zone of E(5°) = 400 Wm^{-2}. The radiant flux density E(O) = E_{sol} + $E_{aur}(\Theta)$ is always lower in presence than in absence of cirrus. Although scattering reduces E(Θ), by increasing the receiver aperture more energy can be collected particularly if cirrus is present.

3.4 Increase of Solar Radiance by Decreasing Θ
It is possible that a reduction in the receiver aperture is advantageous for the operation of a focusing solar high temperature device. Such a gain in the radiance can be obtained by reducing Θ to values smaller than Θ_R. The Sun exhibits a marked limb darkening. It is strongly dependent on the wavelength because of the temperature gradient in the outer layers of the Sun. Fig.3.5 shows the limb darkening measured at several wavelengths. Fig 3.6 shows the limb darkening of the spectrally integrated radiation. Scheffler and Elsässer's (1974) table of the limb darkening of the integrated radiation is repeated here as Table 3.3. The analytical approximation reads

$$L(r)/L(0) = 0.4 \cdot (1 + 1.5 \cdot (1 - r^2)^{1/2}) \qquad (3.20)$$

$r=R/R_S$ is the relative distance from the center of the solar disc with radius R_S. The equation presents the radiant flux density of the inner part of the solar disc up to radius $R = R_S/2$

$$E(R_S/2) = 0.3 \cdot E(R_S) \qquad (3.21)$$

By comparison with the areas of the solar disc

$$A(R_S/2) = 0.25 \cdot A(R_S) \qquad (3.22)$$

we obtain the radiances of the Sun averaged over the corresponding areas of the disc

$$L(R_S/2) = 1.2 \cdot L(R_S) \qquad (3.23)$$

The radiance averaged over the inner part of the Sun's disc appears to be 20 % greater than the radiance averaged over the full disc.

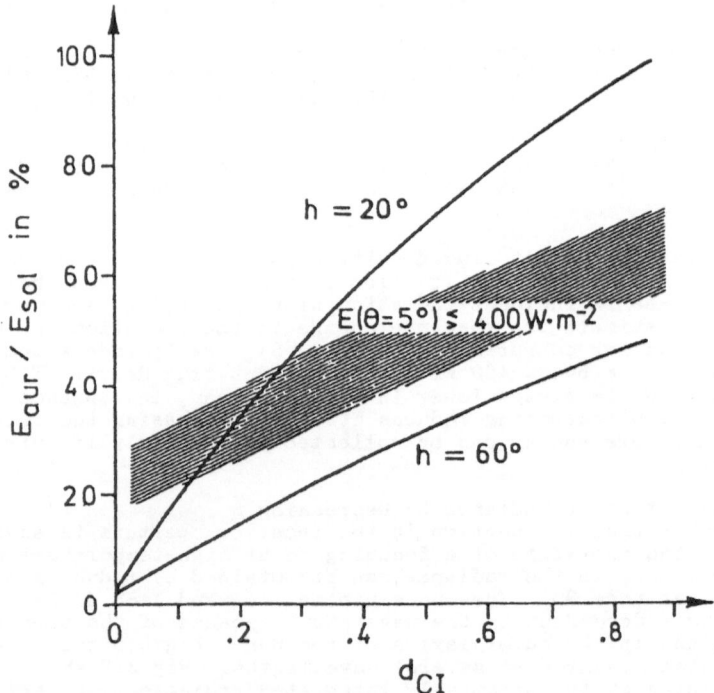

Fig.3.4 Circumsolar Ratio $E_{aur}(\theta=5^{\circ})/E_{sol}$ as function of cirrus optical thickness at two solar elevations h. $d_{AE}=0.14$. The hatched area indicates the approximate boundary for values of $E(\theta=5^{\circ})=400$ Wm^{-2}; values of $E>400$ Wm^{-2} lie above, of <400 Wm^{-2} below the bound

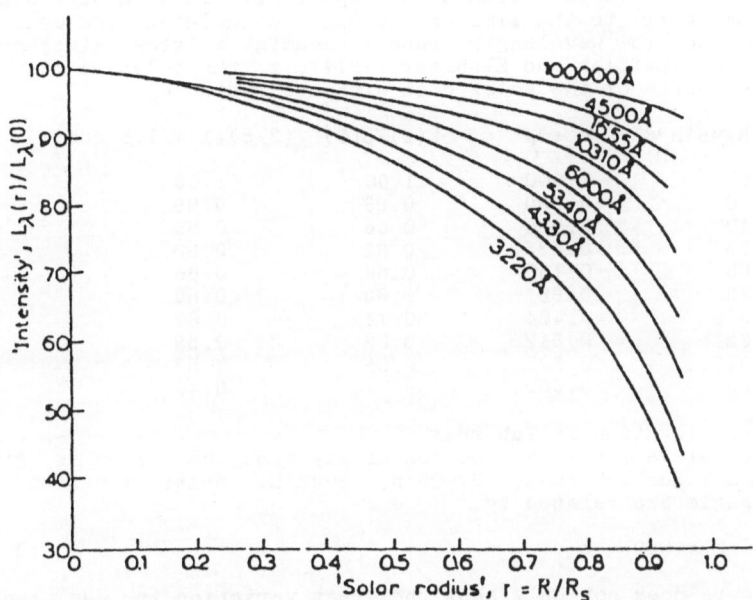

Fig.3.5   Spectral behaviour of the limb darkening of the relative
solar radiance L(R)/L(0). Taken from Scheffler and Elsässer (1974)

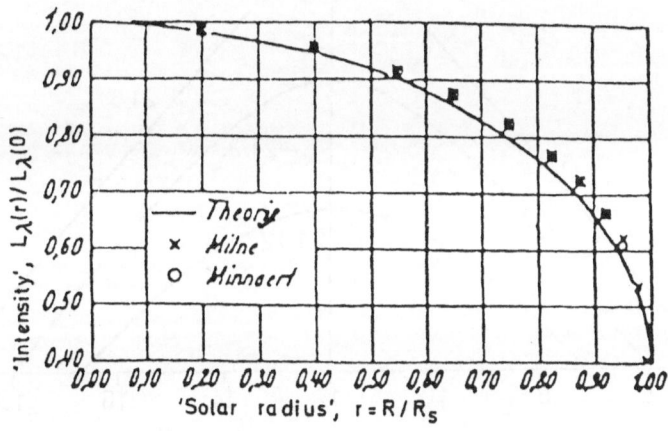

Fig.3.6   Limb darkening of the spectrally integrated relative
solar radiance L(r)/L(0). Reproduced from Waldmeier (1941)

Table 3.3 Relative variation of the spectrally integrated radiance from the centre to the limb of the Sun, calculated from measurements at several wavelengths resp. assuming a 'gray atmophere'. Taken from Scheffler and Elsässer (1974). r, the relative distance from the centre of the solar disc with its radius R_S

$r=R/R_S=\sin\varphi$	$\cos\varphi$	$L(r)/L(0)$	$(2/5)(1 + 1.5\cdot\cos\varphi)$
0.00	1.000	1.00	1.00
0.20	0.980	0.99	0.99
0.40	0.916	0.96	0.95
0.55	0.835	0.92	0.90
0.65	0.760	0.89	0.86
0.75	0.661	0.83	0.80
0.875	0.484	0.74	0.69
0.95	0.312	0.63	0.59
0.975	0.222	0.55	0.53
1.00	0.000	-	0.40

3.5 Solar Elevation at Tabernas

The solar elevation as a function of day time and season is given in Fig.3.7 for Tabernas, 37°06'N, 2°23'E. Solar elevation and zenith angle are related by

$$h = 90° - z \qquad\qquad (3.24)$$

The figure does not only show the great variation in day length over the year but also the variation of the maximum solar elevation which is about 30° in winter and over 76° in summer.

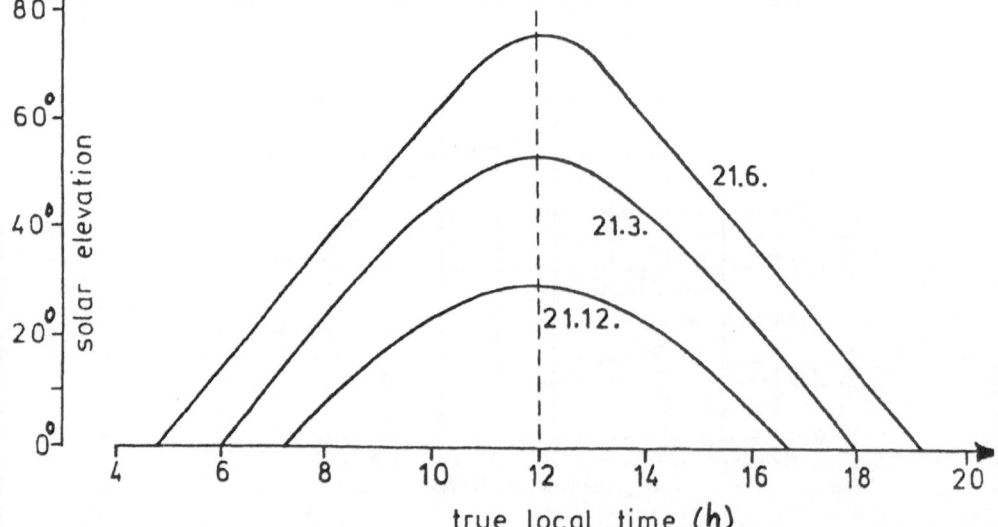

Fig.3.7 Daily course of the solar elevation at Tabernas for solstices and equinoxes

3.6 Meteorologically Relevant Optical Quantities

According to Section 3.2 the relevant optically active meteorological quantities are, in order of their importance
- thick clouds
- cirrus clouds with optical thickness d_{CI}
- aerosol with optical thickness d_{AE} of visual range VIS
- water vapour with optical thickness d_{WA}
- barometric pressure p, which controls the Rayleigh scattering by the air molecules with optical thickness d_{RA}
- ozone with optical thichness d_{O3}
- other gaseous absorbers, with optical thickness d_{GA}.

Cloud coverage is classified by cloud type and degree of cloud cover, indicating the proportion of the sky obscured by clouds. For the present purpose we distinguish between 'stratiform' (horizontally extended clouds) and 'convective' clouds (scattered clouds). Stratiform clouds cast shadows that last a long time and the solar device operation can be steered e.g., with radar or satellite images to predict approaching cloud fields. Convective clouds differ in size distributions with variable and short periods of shadow and sunshine. The radiation available in the gaps between the thick clouds can be calculated accurately enough from the formulae to obtain the strength of the radiation at clear sky.

The standard degree of cloud coverage is because of the Kulisseneffekt not an adequate information for solar usage. A better measure are the 'Cloud Free Line of Sight Probabilities' (Lund et al. 1965, 1978). They indicate the chance that at a given elevation angle cloudless sky can be seen from the ground. This measure, called CFLOSG (G stands for ground), is independent of the azimuth. The isolines of CFLOSG are plotted on geographic maps with a spacing of ΔCFLOSG = 0.05. They are available for four times per day, for the four seasons and at elevation angles of 10°, 30° and 90° (Lund 1978).

The nearest station to Tabernas at which such data were collected is Granada, $37^o08'N$ $3^o37'W$, i.e., about 100 km of distance. These CFLOSG data do not contain information about possible orographic clouds at Tabernas, which can only be included by cloud observation for Tabernas itself, either from the ground or from satellites.

The CFLOSG indicate whether a line of sight is cloud free. Visible cirrus counts as cloud. However, we have treated in Eq.3.2 cirrus clouds separately from other clouds. Therefore, the contribution of cirrus to CFLOSG must be removed when determining P_C. The probability that there are cirrus clouds between the Sun and the concentrating solar device is derived from observations (by looking upwards) from high flying (7.5 km) aircraft (Bertoni,1977).

Observations are available of 'Cloud Free Line of Sight Probabilities' CFLOSA (A for aircraft) and of 'Clear Line of Sight Probabilities', CLOSA. Both values are related mainly to cirrus since the chance of encountering a cumulo-nimbus is very low and no other clouds normally occur in that height.

We have to define the classes of clouds for application in Eq.3.15. To distinguish between CFLOSA and CLOSA and between cirrus and other clouds limits must be set to the various optical thicknesses. This is a matter of judgement. We used as an upper limit the optical thickness 2 and its fractions of 10

$$2 > d_{CI}(1) > 0.2 > d_{CI}(2) > 0.02 > d_{CI}(3) > 0 = d_{CI}(4) \qquad (3.28)$$

The optical thickness of cirrus can exceed 2 but usually it is much less (e.g., Heymsfeld, 1975). The limit 2 for the borderline of cirrus and clouds ($d_C > 2 > d_{CI}$) is chosen because the transmittance of a cloud with $d_C = 2$ is already smaller than 2 %.

The limit $d_{CI} = 0.2$ is chosen as the boundary between 'visible'and 'subvisible'cirrus. Cirrus can be recognized from aircraft by a brightening of the sky or by a colour change towards white or by the contrast between structured clouds and the cloud free sky. The value $d_{CI} = 0.2$ causes the albedo of the cirrus cloud to be 0.02 (Wendling, 1980), a value that corresponds to the smallest degree of contrast perceptible to the human eye. This was the reason for choosing this value, even though under particular conditions, a pilot's ability to spot cirrus starts at different optical thicknesses. When observing well lit cirrus against the sky it starts at noticeably lower values. But it can start at thicknesses higher than 0.2 when observing homogeneous and barely structured cirrus.

We complement the cloud data with the probabilities for cirrus P_{CIS}, extracted from SAGE measurements (Woodbury and McCormick, 1986). The SAGE measurements in occultation against the Sun were intended to measure stratospheric aerosol and trace gases. The resulting data exhibit two features: they are averaged over horizontal distances of several hundreds of kilometers; the optical density of cirrus counted as cloud may be as low as 0.008 km^{-1}. It is not possible to decide whether these values are derived from a short thick cirrus cloud or from a long thin one. But since it is not necessary to know the probability of the presence of cirrus $P_{CI}(i)$ separately from the transmittance $t_{CI}(i)$ but only the sum of products $P_{CI}(i) \cdot t_{CI}(i)$ (see Eq.3.15) above the location, it does not matter whether SAGE has measured a short thick or a long thin cirrus. Cirrus may extend vertically for 100 or 1000 m. The SAGE measurements report on cirrus down to $d_{CI} = 0.002$ or less. Thus, they cover part of the range which is 'clear' to the eye. The chosen limits of $d_{CI} = 0.02$ corresponds to a transmittance of 0.98 along the vertical path and, therefore, it cannot any more be detected in transmission by an observer. The vertical thickness of cirrus clouds is small compared to their horizontal extension: they do not show a Kulisseneffect. Therefore it is allowed to use averages of $P_{CI}(i)$ observed at 30° and 60° elevation. This is of advantage, since rather few air craft observations are available. Thick water clouds and cirrus clouds can be treated independently. Thus, the Cloud Free Line of Sight Probabilities from ground are related to P_C and $P_{CI}(1)$ as

$$CFLOSG = (1 - P_C)(1 - P_{CI}(1)).$$

Cirrus clouds of different optical depths are not existing simultaneously. Thus, the following equations relate the Lines of Sight Probabilities to the $P_{CI}(i)$

$$CFLOSA = 1 - P_{CI}(1) \qquad CLOSA = 1 - (P_{CI}(1) + P_{CI}(2))$$

$$P_{CIS} = P_{CI}(1) + P_{CI}(2) + P_{CI}(3) \tag{3.26}$$

It follows

$$P_{CI}(1) = 1 - CFLOSA \qquad P_{CI}(2) = CFLOSA - CLOSA \tag{3.27}$$

$$P_{CI}(3) = P_{CIS} + CLOSA - 1 \qquad P_{CI}(4) = 1 - \sum_{i=1}^{3} P_{CI}(i)$$

The $P_{CI}(i)$ corresponding to the available data were determined for the four seasons. Spring, summer, autumn and winter are presented in Table 3.4 and Fig. 3.8. The annual cycle of both thin and thick cirrus exhibits a minimum in summer; thick cirrus shows a maximum in winter; the probability of thick cirrus is small.

Table 3.4 Probability of the presence of cirrus clouds with several optical thicknesses $d_{CI}(i)$ over Tabernas. The values of $d_{CI}(1)$, $d_{CI}(2)$, and $d_{CI}(3)$ are the geometric means of the limits of the classes 2 - 0.2; 0.2 - 0.02; 0.02 - 0.002

		i = 1	2	3	4
	$d_{CI}(i)$ =	0.6	0.06	0.006	0.0
Spring	MAM	0.050	0.115	0.435	0.4
Summer	JJA	0.055	0.02	0.025	0.9
Autumn	SON	0.050	0.03	0.220	0.7
Winter	DJF	0.090	0.085	0.125	0.7

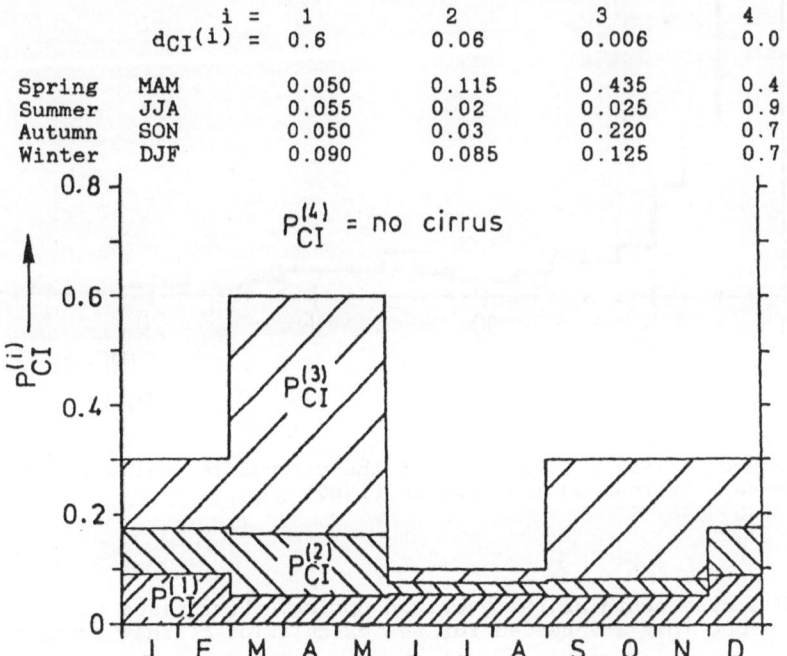

Fig.3.8 Probability of the occurence of cirrus P_{CI} in the seasons with the optical thickness of cirrus put into four classes

The behaviour of cirrus during the seasons is confirmed by the values of the circumsolar ratios which were measured in Tabernas for $E_{sol} > 400$ Wm^{-2} (Grasse, 1985). Fig. 3.9 of the SSPS report (Grasse, 1985) shows the frequency distribution of the circumsolar ratio from summer through autumn 1983. The circumsolar ratio is understood as the quantity E_{aur}/E_{sol} according to Eq. 3.19 for an angular width of the aureole of $\Theta = 5°$, so that it can be compared with Fig. 3.4. These measurements show that thick cirrus is indeed very rare in Tabernas. Circumsolar ratios greater than 5%, which occur more frequently, can be caused either by thin cirrus in classes CI(2) and CI(3), (which are normally not recognized as cirrus by observers) or by optically thick aerosols.

NUMBER OF MEASUREMENTS

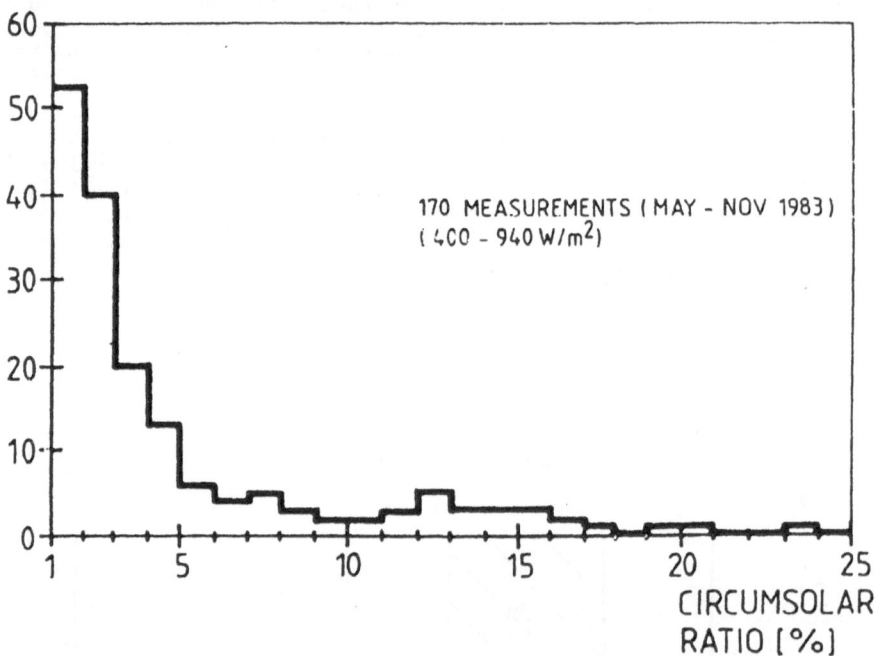

Fig.3.9 Probability distribution of the circumsolar ratio measured at Tabernas. Reproduced from Grasse (1985)

Table 3.4 also contains the averaged optical thicknesses of cirrus $d_{CI}(i)$ since they are needed for the calculation of the transmit-
0.002

$$d_{CImin}(i) = (d_{CImin}(i) \cdot d_{CImax}(i))1/2 \qquad (3.28)$$

- 34 -

The values of the optical thicknesses are then 0.6 and fractions of 10.

The probability of thick clouds follows from Eq.3.26

$$P_C = (CFLOSA - CFLOSG)/CFLOSA \qquad (3.29)$$

The CFLOSA values are only available as seasonal averages, but they do not vary diurnally or show a Kulisseneffekt. So the diurnal variation and dependence on the elevation present in CFLOSG is carried over into P_C.

Table 3.5 and Fig. 3.10 show P_C determined in this way as a function of the true local time. Observations are available only at the times where symbols are plotted; values in between are linearly interpolated. P_C is higher in winter than in summer (sunny summers in Spain), higher for lower than for higher elevations. P_C shows an increase in the cloudiness in the afternoon (except in summer), which is also reported for Tabernas by Grasse (1985, p.127).

Table 3.5 Probability P_C of the presence of thick clouds over Tabernas, depending on the season, hour of the day and elevation h of the CFLOSG

Season	Hour of the day in true local time	h = 10°	h = 30°	h = 90°
Spring MAM	1	0.35	0.27	0.20
	7	0.50	0.40	0.36
	13	0.53	0.45	0.40
	19	0.56	0.45	0.33
Summer JJA	1	0.18	0.12	0.10
	7	0.25	0.19	0.13
	13	0.20	0.15	0.07
	19	0.24	0.15	0.09
Autumn SON	1	0.37	0.28	0.23
	7	0.48	0.37	0.35
	13	0.50	0.38	0.31
	19	0.43	0.36	0.30
Winter DJF	1	0.24	0.19	0.13
	7	0.36	0.29	0.24
	13	0.46	0.35	0.29
	19	0.45	0.35	0.30

Since at a particular location and time the Sun has a fixed elevation (Fig.3.7), the values of P_C that are valid for line of sight towards the Sun must be taken from Fig.3.10.

Fig.3.11 shows the diurnal variation of P_C obtained in this way for solar elevations which differ in 10°. The values for sunset and sunrise are extrapolated assuming CFLOSG(0°) = 0.5·CFLOSG(10°) because the probability of a cloud free line of sight steeply drops towards the horizon. Sunsets and sunrises can still be occasionally observed. The quality of this extrapolation does not really matter since in solar energy usage the device will certainly not be in operation when the solar elevation is lower than 10°. The variation in the strength of the incident radiation during the day is mostly controlled by the changing elevation of the Sun. The diurnal variation of cloudiness is only of secundary importance. In particular it is obvious that there are not many clouds in summer.

To determine the annual variation in P_C, further data must be used. Lund et al. (1978) provide a matrix from which the cloud cover in tenths can be converted into P_C values as a function of the elevation. Unfortunately we cannot use this approach as the data needed (cloud coverage) are not available. Instead, values for P_C are read from Fig.3.11 which hold for the middle part of the day, where the solar elevations are relatively high; we regard them as valid for the day as a whole. So we arrive at one P_C value for each of the four seasons:

Spring	$P_C = 0.42$
Summer	$P_C = 0.14$
Autumn	$P_C = 0.38$
Winter	$P_C = 0.35$

To derive the annual variation, the P_C are plotted in Fig.3.12 as solid squares together with the annual variation of the cloud cover over Tabernas from Malberg (1977). Other sources for the cloud cover (e.g., Köppen and Geiger, 1932) show similar annual variations for places close by, although they do show different absolute values and do not show the relative minimum in October. This indicates that the annual variation of the cloud coverage given in Fig.3.12 properly represents the facts and can be used as the annual variation of P_C. These P_C values are listed in Table 3.6. They are in accordance with Grasse (1985, p.127), who reports that in Tabernas the months February, March and November are rather cloudy.

There are not many data available on the atmospheric turbidity. One could use measurements or observations of the visibility range from airports. But we prefer to use values of the optical thickness which were measured at some of the WMO's BAPMoN (Background Atmospheric Polution Monitoring Network) stations. Spain has only one such BAPMoN station. In 1974 till 1977 it has been Torrejon. We will use its recordings here, because Torrejon (40°29'N 3°28'W) lies at 611 m above sea level and has a climate similar to Tabernas: dry and dusty summers and a reduction in the dust load in autumn after rains, when the ground shows again vegetation. Fig.3.13 gives the mean monthly optical thickness at $\lambda = 0.5$ μm. The BAPMoN data are averaged and converted into d_{AE} according to Eq.3.14. These values are given in Table 3.6.

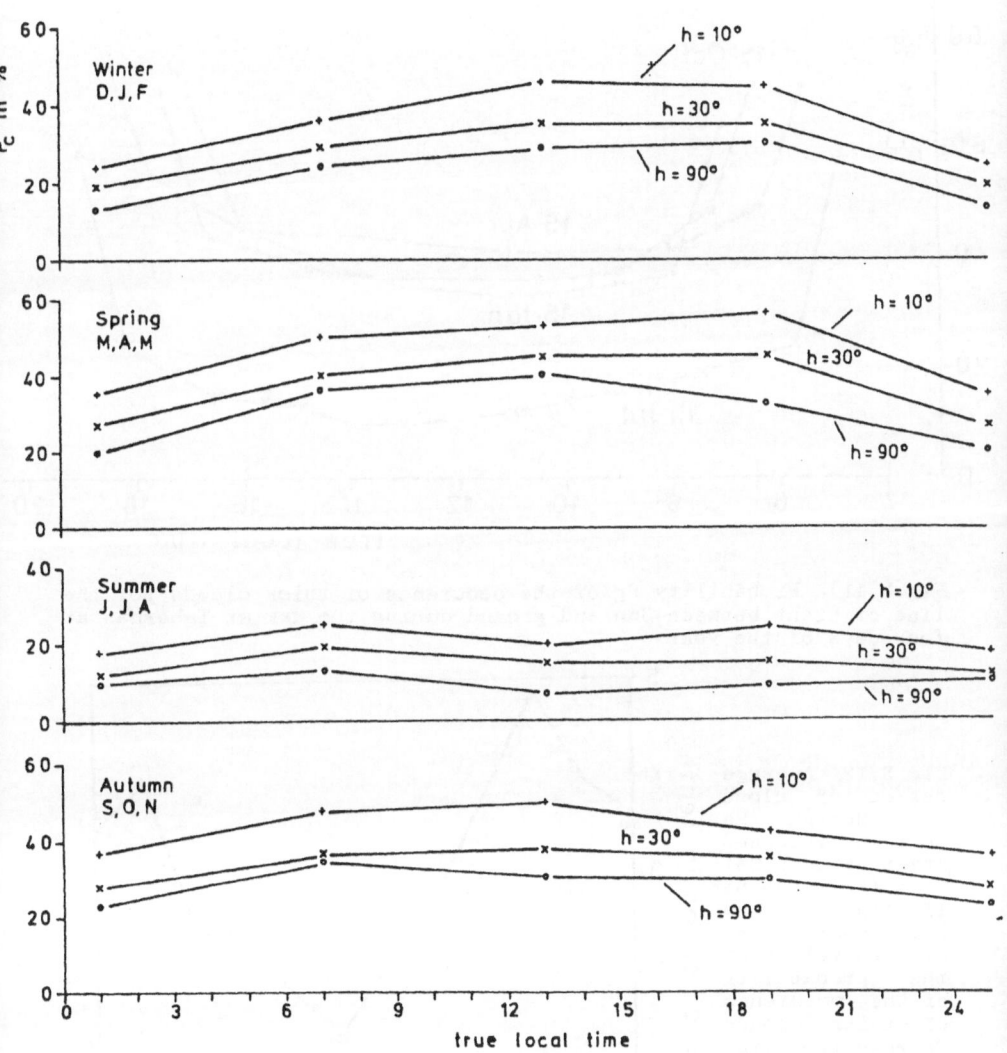

Fig.3.10   Probability of the occurence of thick clouds P_C during the day at Tabernas for the four seasons and three elevations h of observation directions, independent of the azimuth

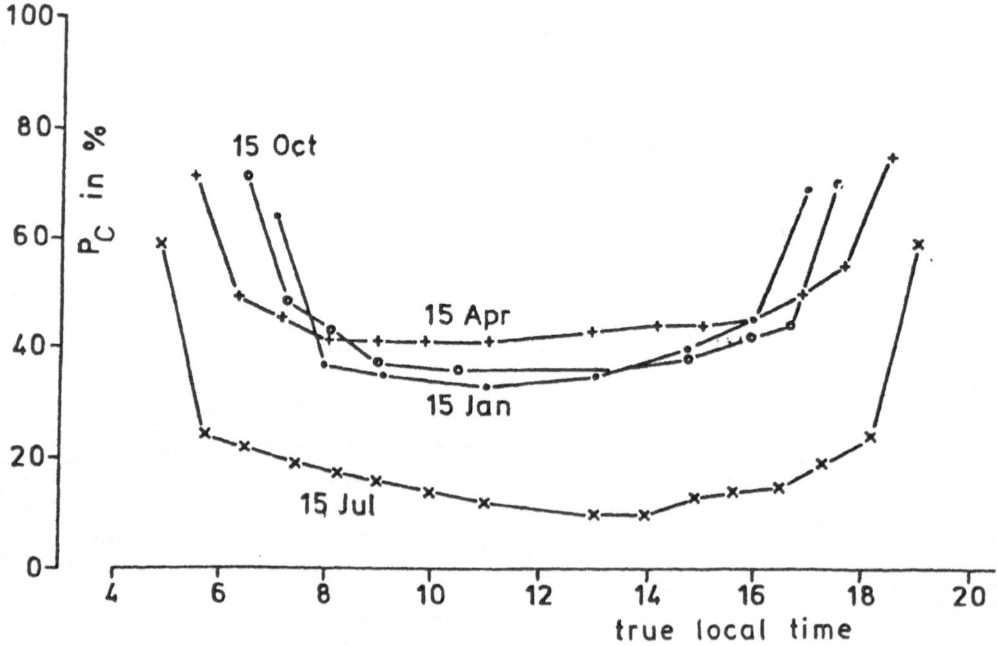

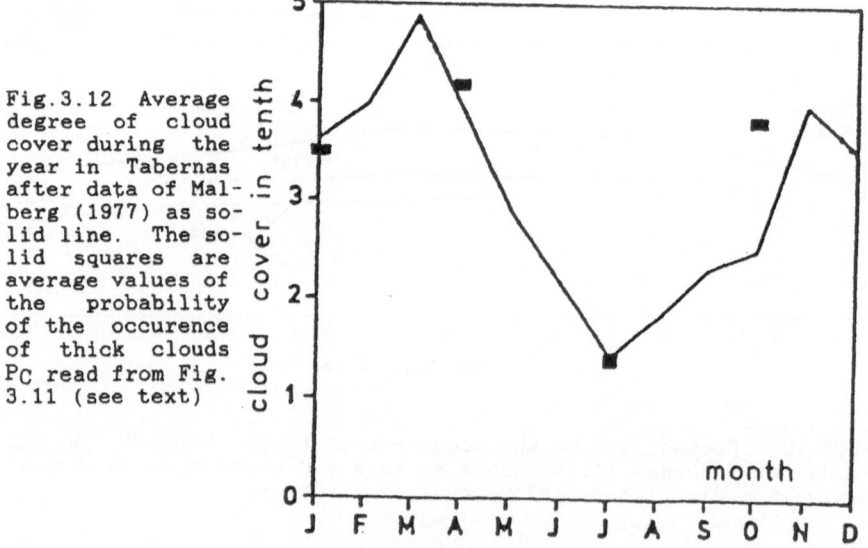

Fig.3.11 Probability P_C of the occurence of thick clouds for the line of sight between Sun and ground during the day at Tabernas at four days of the year

Fig.3.12 Average degree of cloud cover during the year in Tabernas after data of Malberg (1977) as solid line. The solid squares are average values of the probability of the occurence of thick clouds P_C read from Fig. 3.11 (see text)

Table 3.6 Variation of several atmospheric quantities valid for Tabernas over the year: probability P_C of the presence of thick clouds; optical thickness of the aerosol particles d_{AE}; content of water vapour in the vertical column of the atmosphere u_{WA} and of ozone u_{O3}; the distance factor f_3 (Eq 3.3) between Sun and Earth

Month	P_C	d_{AE}	u_{WA} $g \cdot cm^2$	u_{O3} cm(STP)	f_3
January	0.35	0.128	1.33	0.28	1.032
February	0.40	0.144	1.33	0.29	1.024
March	0.48	0.160	1.33	0.30	1.010
April	0.42	0.220	1.81	0.31	0.993
May	0.29	0.238	1.81	0.30	0.977
June	0.21	0.288	2.29	0.29	0.968
July	0.14	0.259	2.41	0.27	0.967
August	0.18	0.283	2.65	0.26	0.975
September	0.24	0.214	2.41	0.25	0.987
October	0.38	0.147	2.41	0.25	1.007
November	0.40	0.125	1.81	0.25	1.022
December	0.35	0.097	1.45	0.27	1.032

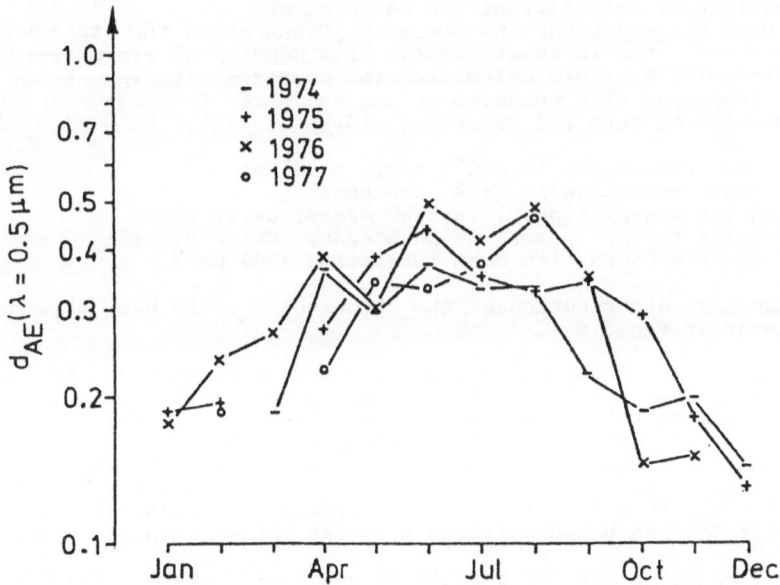

Fig.3.13 Monthly mean averages of the spectral aerosol optical depth d_{AE} at wavelength 0.5 um recorded at the BAPMoN station at Torrejon in the years 1974 through 1977

There are measurements which can be used for comparison with the present calculated values of Section 3.7. Since the maximum values in the measurements occur in optically thin atmospheres, i.e., where the aerosol load is slight, values of the optical thickness are required under those conditions. The minimum values of the optical thickness d_{AE} from the four years for which measurements are available show the annual variation with a maximum in summer. The minimum monthly d_{AE} are, however, so small that for the minimum relative air mass the aerosol transmittance t_{AE} is at least 0.85.

The amount of water vapour over Tabernas is read from maps of Tuller (1968). The figures converted to $g \cdot cm^{-2}$ and corrected after Iqbal (1983) to the barometric pressure at Tabernas with $u_{WA}(p) = u_{WA}(p_0) \cdot (p/p_0)^{3/4}$ are presented in Table 3.6. The maximum in summer is caused by the water vapour saturation rising with temperature.

In Table 3.6 ozone is given from Gebhardt et al. (1970). The annual variation in the factor f_3 as included in Table 3.6 is taken from Möller (1970). To calculate the transmittance due to scattering by air molecules and the absorption caused by gases with constant mixing ratio requires only the barometric pressure p to weight the relative air mass. This is in Tabernas 950 hPa on the average, equivalent to a height of 500 m above sea level.

3.7 Comparison of Calculations and Measurements
To validate the equations and the assumptions about the atmospheric parameters, the incident radiant flux density at ground level and other solar data are calculated and compared with values measured at Tabernas. The measured values are taken from the report SSPS, 'Results of test and operation' (Grasse, 1985). We have used

> Beam insolation in Wm^{-2}, maximum values
> Beam insolation in Wm^{-2}, averages
> Solar energy (power) in $kWhm^{-2}$/day, daily peak
> Solar energy (power) in $kWhm^{-2}$/day, daily average
> Monthly hours with beam insolation >300 Wm^{-2}.

The frequency distribution of the circumsolar ratio has already been treated in Fig.3.9.

The measurements of the solar radiation were made with an Eppley Pyrheliometer which includes the aureole up to $\theta = 5^{\circ}$. This quantity is by Grasse (1985) called 'beam insolation'. In the present notation it is $E(\theta = 5^{\circ}) = E_{sol} + E_{aur}(\theta = 5^{\circ})$. In order to compare the calculations with Grasse's measurements, the values of E_{sol} are treated according to Eq.3.1. They are multiplied by the factor $(1 + E_{aur}(\theta = 5^{\circ})/E_{sol})$. The E_{aur}/E_{sol} are constructed from Fig.3.4 for d_{CI} and h and weighted with the $P_{CI}(i)$ from Table 3.4.

Table 3.7 contains all the quantities mentioned above. Examples are calculated for January and July, based both on E_{sol} and $E(0 = 5^{\circ})$ for comparison with the measured values of the beam insolation. This again shows the additional energy gained by including the aureole.

Table 3.7 Quantities of solar radiation calculated for Tabernas, using both direct solar radiation E_{sol} and direct plus scattered radiation $E_{sol} + E_{aur}(\Theta = 5°)$ for comparison with the measurements (Figs.3.14, 3.15 and 3.16). More explanation is given in the text

	January		July	
	direct sun	direct sun + aureole	direct sun	direct sun + aureole
Maximum value of radiant flux density, Wm^{-2}	921	921	925	925
Mean value of radiant flux density Wm^{-2}	354	387	500	523
Mean value of solar radiation $kWhm^{-2}$ per day	3.2	3.4	6.6	7.0
Available solar radiation when $E > 300\ Wm^{-2}$ in hours per month	196	209	338	343

The maximum values of the measured beam insolation coincide with the beam insolation calculated from the minimum measured optical thickness, the minimum relative air mass and no cirrus. Under these conditions is the increase in the radiant flux density when including the aureole not significant. The incident radiation in January and July is nearly the same which surprises at first glance. It happens that the effect of the smaller relative air mass in summer is counterbalanced by the simultaneous increase in the values for water vapour and turbidity. In Fig.3.14 the maximum values of E_{sol} are inserted as the upper solid bars; otherwise the figure is reproduced from the SSPS report (Grasse, 1985). The measured and the calculated values match well.

The other quantities in Table 3.7 are calculated with the diurnal variation of the incident radiation. They are calculated from step functions with n steps $h^* = 10°·n$, where each particular solar elevation h^* is valid for its period Δt when $h = h^* \pm 5°$. In the mean values both cirrus and thick clouds are included. The diurnal variation of the thick clouds is taken from Fig.3.11.

The mean values of the calculated beam insolation have been inserted into Fig.3.14 as the lower solid bars. They also match well although the values for January are a little low. The reason may be the uncertainty in the probabilities P_C of the thick clouds.

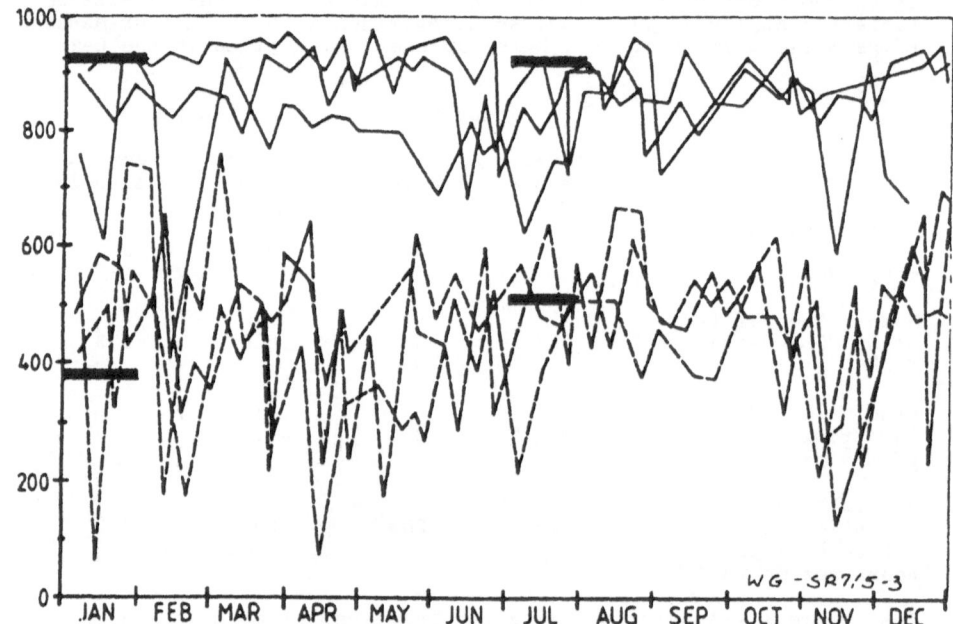

INSOLATION [W/m²] ——— MAX. VALUES ------ WEEKLY AVERAGES

Fig.3.14 Maximum values and weekly averages of the 'beam insolation' E_{sol} measured in Tabernas in the years 1982 through 1984,
reproduced from Grasse (1985). The inserted solid horizontal bars
are calculated from the atmospheric data used in this investigation. The upper bars belong to atmospheres with low aerosol load
and without clouds; the lower bars belong to mean aerosol load and
average cloudiness (see text)

The present calculated mean values of the solar energy in kWhm⁻²
per day have been inserted as solid crosses to the measured values
(Grasse, 1985) in Fig.3.15. They are based on the same diurnal
variations as the radiation flux densities. The match with the
measurements is about as close as in Fig.3.14.

The relationship is not as close for the number of hours with beam
insolation $E(0 = 5^o) > 300$ Wm⁻². In this case Fig.3.16 shows that
the calculated values are larger than the measured values, both
for January and July. The day lengths are always taken on the 15th
of the month. If we compensate for the variation of day length,
the calculated values would be little lower in July and little
higher in January. There is another uncertainty in the calculation. The diurnal variation is represented only roughly by step
functions from which, although they have been smoothed, the value
of 300 Wm⁻² can only be read with an accuracy of + 0.1 hours. Over
a month the uncertainty in the number of hours with $E(0 = 5^o)$
> 300 Wm⁻² can add up to more than 12 hours.

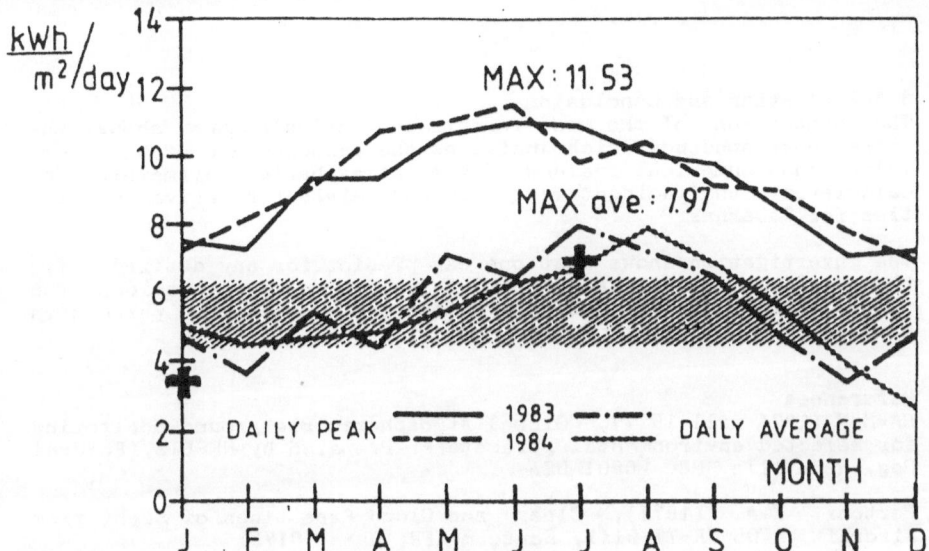

Fig.3.15   Monthly averages of incident solar energy per day in Tabernas, 1983 and 1984, taken from Grasse (1985). The inserted solid crosses are calculated from atmospheric data for average conditions including clouds

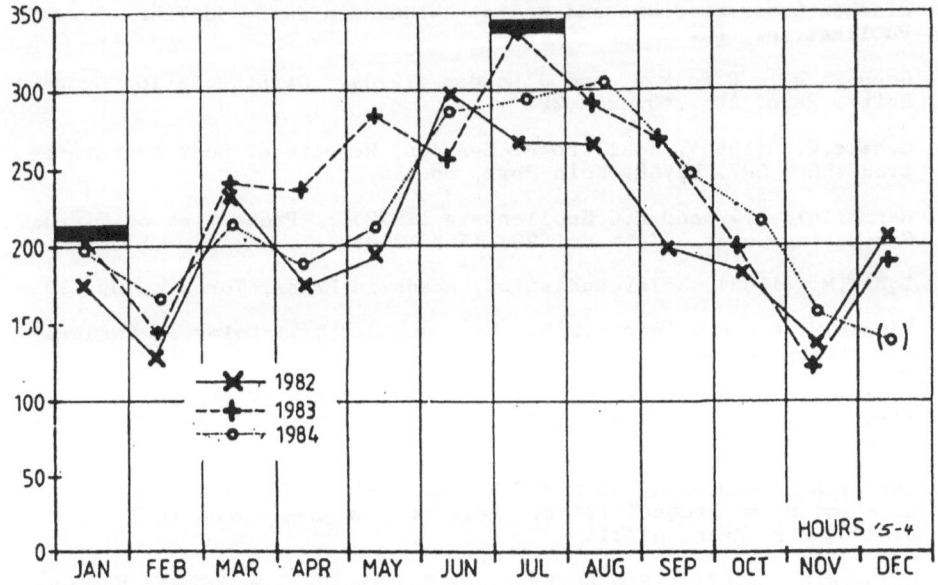

Fig.3.16   Hours per month with beam insolation $E_{sol} > 300$ Wm^{-2} in Tabernas, 1982 through 1984, reproduced from Grasse (1985). The inserted solid horizontal bars are calculated from the data used in this investigation for average conditions including clouds

3.8 Evaluation and Conclusion

The comparison of the measurements and calculations shows the close correspndence which justifies the proposed use of the formulae and numerical values of the atmospheric parameters for calculating the incident radiation and related radiative quantities for Tabernas.

The investigation shows that one can predict for any desired site the incident direct radiation with good accuracy, provided the values of the relevant optically effective atmospheric parameters are known to the accuracy required.

References

BAPMoN (1974 till 1977), Global Atmospheric Background Monitoring for selected environmental parameters. Prepared by NESDIS, Federal Bdg, Asheville, NC. 28801 USA

Bertoni, E.A. (1977), Clear and Cloud Free Lines of Sight from Aircraft. AFGL-TR-77-0141, Hanscom AFB, Mass. 01731

Bird,R. and R.L.Hulstrom (1981), Review, evaluation and improvement of direct irradiance models, Trans. ASME. J.Sol.Energy Eng. 103, 182-192

Fröhlich,K. (1985), Jahresbericht 1984 des Physikalisch-Meteorologischen Observatoriums und Weltstrahlungszentrum, CH-7260 Davos; Publikat. Nr. 613

Gebhart,R., R.Bojkov and J.London (1970), Stratospheric Ozone, Beitr. Phys. Atm.,43, 209-227

Grasse,W. (1985), Small Power Systems; Results of test and operation. SSPS SR7. DFVLR, Köln Porz, pp.155

Heymsfield,A.J. and R.G.Knollenberg (1972), Properties of Cirrus Generating Cells. J.Atm.Sci.29, 1358-1366

Iqbal,M. (1983), Solar Radiation, Academic Press, Toronto, pp.390

Köppen,W. and R.Geiger (1936), Handbuch der Klimatologie, Borntrager Verlag, Berlin

Lund,I.A. (1965), Estimating the probability of clear lines of sight from sunshine and cloud cover observations. J.Appl.Meteorology, 4, 714-722

Lund,I.A., D.D.Grantham and C.B.Elam (1978), Atlas of Cloud Free Line of Sight Probabilities, Part 4: Europe, AFGL-TR-78-0276, Hanscom AFB, Mass. 01731

Malberg,H. (1977). Ein Beitrag zur Bewölkungsklimatologie Europas und des Atlantiks anhand von Satellitenaufnahmen. Meteorologische Abhandlungen, Inst.f.Meteorologie der FU Berlin, Neue Folge, Serie A, Bd.1, Heft 1

Möller,F. (1970) in: Meteorologisches Taschenbuch, F.Baur (Ed.), Akademische Verlagsgesellschaft Leipzig
Scheffler,H. und H.Elsässer (1974), Physik der Sterne und der Sonne. Wissenschaftsverlag Mannheim, pp.535

Thomalla,E., P.Koepke, H.Müller und H.Quenzel (1983), Circumsolar radiation calculated for various atmospheric conditions. Solar Energy 30, 575-587

Tuller,S.E. (1968), World distribution of mean monthly and annual precipitable water. Month. Weath. Rev. 96,785-797

Waldmeier,M. (1941), Ergebnisse und Probleme der Sonnenforschung. Akademische Verlagsgesellschaft Leipzig, pp. 264

Wendling,P. (1980), On the albedo and infrared emissivity of cirrus clouds. Int.Rad.Symp. 1980, Mt.Collins, CO, USA

Woodbury,G.E. and M-P.McGormick (1986), Zonal and geographical distributions of cirrus clouds determined from SAGE data. J.Geoph.Res. 91, noc C2, 2775-2785

4. Numerical results

According to the recipes discussed in Section 2 with the use of the meteorological input of Section 3 numerical results are computed and listed in the Table 4.1. The graphical presentations are collected in the following order

- Available Process Heat ($kWh/m^2/d$) for Clear Sky and Cloudy Sky Monthly Averages, Summarized in a Yearly Average
- Maximum Operation Time (hours/d) for Clear Sky and Cloudy Sky Monthly Averages, Summarized in a Yearly Average
- Efficiency (Process Heat Obtained/Solar Radiative Input) Monthly Averages, Summarized in a Yearly Average
- Available Solar Energy ($kWh/m^2/d$) for Clear Sky and Cloudy Sky Monthly Averages, Summarized in a Yearly Average

In all the diagrams the independent variable is the reduced loss coefficient V as discussed in detail in Section 2.

Losses kWh/m2/d	Avail.Heat no clouds kWh/m2/d	Avail.Heat + clouds kWh/m2/d	Efficiency	Opera.Time no clouds hours/d	Opera.Time + clouds hours/d	Avail.Sun no clouds kWh/m2/d	Avail.Sun + clouds kWh/m2/d	
800	0	0	0	0	0	4.423	2.874	January
750	0	0	0	0	0	4.423	2.874	January
700	0	0	0	0	0	4.423	2.874	January
650	.001	0	0	.243	.157	4.423	2.874	January
600	.067	.043	.014	2.668	1.734	4.423	2.874	January
550	.242	.157	.054	4.217	2.741	4.423	2.874	January
500	.479	.311	.107	5.239	3.405	4.423	2.874	January
450	.76	.493	.171	6.023	3.914	4.423	2.874	January
400	1.077	.7	.242	6.663	4.33	4.423	2.874	January
350	1.423	.924	.321	7.206	4.683	4.423	2.874	January
300	1.795	1.166	.405	7.68	4.991	4.423	2.874	January
250	2.188	1.422	.494	8.100001	5.265	4.423	2.874	January
200	2.602	1.691	.587	8.477	5.51	4.423	2.874	January
150	3.034	1.972	.685	8.819	5.732	4.423	2.874	January
100	3.482	2.263	.787	9.137	5.939	4.423	2.874	January
50	3.945	2.564	.891	9.429	6.128	4.423	2.874	January
0	4.423	2.874	1	9.704001	6.307	4.423	2.874	January

Losses kWh/m2/d	Avail.Heat no clouds kWh/m2/d	Avail.Heat + clouds kWh/m2/d	Efficiency	Opera.Time no clouds hours/d	Opera.Time + clouds hours/d	Avail.Sun no clouds kWh/m2/d	Avail.Sun + clouds kWh/m2/d	
800	0	0	0	0	0	5.183	3.109	February
750	0	0	0	0	0	5.183	3.109	February
700	.002	.001	0	.348	.208	5.183	3.109	February
650	.069	.041	.012	2.578	1.546	5.183	3.109	February
600	.247	.148	.046	4.348	2.608	5.183	3.109	February
550	.493	.295	9.399999E-02	5.465	3.279	5.183	3.109	February
500	.787	.472	.15	6.318	3.79	5.183	3.109	February
450	1.12	.672	.215	7.014	4.208	5.183	3.109	February
400	1.485	.891	.285	7.604	4.562	5.183	3.109	February
350	1.878	1.126	.361	8.112001	4.867	5.183	3.109	February
300	2.294	1.376	.441	8.57	5.142	5.183	3.109	February
250	2.732	1.639	.526	8.975001	5.385	5.183	3.109	February
200	3.19	1.914	.614	9.345	5.607	5.183	3.109	February
150	3.665	2.199	.706	9.687	5.812	5.183	3.109	February
100	4.156	2.493	.801	10.003	6.001	5.183	3.109	February
50	4.663	2.797	.899	10.291	6.174	5.183	3.109	February
0	5.183	3.109	1	10.566	6.339	5.183	3.109	February

Losses kWh/m2/d	Avail.Heat no clouds kWh/m2/d	Avail.Heat + clouds kWh/m2/d	Efficiency	Opera.Time no clouds hours/d	Opera.Time + clouds hours/d	Avail.Sun no clouds kWh/m2/d	Avail.Sun + clouds kWh/m2/d	
800	0	0	0	0	0	6.267	3.258	March
750	.006	.003	0	.73	.379	6.267	3.258	March
700	.11	.057	.017	3.392	1.763	6.267	3.258	March
650	.321	.166	.05	4.965	2.581	6.267	3.258	March
600	.597	.31	9.399999E-02	6.072	3.157	6.267	3.258	March
550	.923	.479	.146	6.946	3.611	6.267	3.258	March
500	1.288	.669	.204	7.664	3.985	6.267	3.258	March
450	1.686	.876	.268	8.278	4.304	6.267	3.258	March
400	2.113	1.098	.336	8.817001	4.584	6.267	3.258	March
350	2.565	1.333	.408	9.295	4.833	6.267	3.258	March
300	3.04	1.58	.484	9.722	5.055	6.267	3.258	March
250	3.535	1.838	.563	10.116	5.26	6.267	3.258	March
200	4.049	2.105	.645	10.477	5.448	6.267	3.258	March
150	4.58	2.381	.73	10.808	5.62	6.267	3.258	March
100	5.128	2.666	.818	11.115	5.779	6.267	3.258	March
50	5.69	2.958	.907	11.404	5.93	6.267	3.258	March
0	6.267	3.258	1	11.68	6.073	6.267	3.258	March

Losses kWh/m2/d	Avail.Heat no clouds kWh/m2/d	Avail.Heat + clouds kWh/m2/d	Efficiency	Opera.Time no clouds hours/d	Opera.Time + clouds hours/d	Avail.Sun no clouds kWh/m2/d	Avail.Sun + clouds kWh/m2/d	
800	0	0	0	0	0	6.597	3.826	April
750	0	0	0	0	0	6.597	3.826	April
700	.072	.041	.01	3.111	1.804	6.597	3.826	April
650	.277	.16	.041	4.959	2.876	6.597	3.826	April
600	.557	.323	.084	6.203	3.597	6.597	3.826	April
550	.891	.516	.134	7.167	4.156	6.597	3.826	April
500	1.27	.736	.192	7.97	4.622	6.597	3.826	April
450	1.685	.977	.255	8.667	5.026	6.597	3.826	April
400	2.133	1.237	.323	9.288	5.387	6.597	3.826	April
350	2.611	1.514	.395	9.833999	5.703	6.597	3.826	April
300	3.115	1.806	.472	10.35	6.003	6.597	3.826	April
250	3.643	2.112	.552	10.813	6.271	6.597	3.826	April
200	4.195	2.433	.635	11.248	6.523	6.597	3.826	April
150	4.767	2.764	.722	11.662	6.763	6.597	3.826	April
100	5.358	3.107	.812	12.046	6.988	6.597	3.826	April
50	5.969	3.462	.904	12.406	7.195	6.597	3.826	April
0	6.597	3.826	1	12.752	7.396	6.597	3.826	April

Losses kWh/m2/d	Avail.Heat no clouds kWh/m2/d	Avail.Heat + clouds kWh/m2/d	Efficiency	Opera.Time no clouds hours/d	Opera.Time + clouds hours/d	Avail.Sun no clouds kWh/m2/d	Avail.Sun + clouds kWh/m2/d	
800	0	0	0	0	0	7.086	5.031	May
750	0	0	0	0	0	7.086	5.031	May
700	.095	.067	.013	3.601	2.556	7.086	5.031	May
650	.323	.229	.045	5.417	3.846	7.086	5.031	May
600	.626	.444	.088	6.672	4.737	7.086	5.031	May
550	.984	.698	.138	7.67	5.445	7.086	5.031	May
500	1.389	.986	.195	8.512	6.043	7.086	5.031	May
450	1.832	1.3	.258	9.248	6.566	7.086	5.031	May
400	2.311	1.64	.326	9.904	7.031	7.086	5.031	May
350	2.82	2.002	.397	10.501	7.455	7.086	5.031	May
300	3.358	2.384	.473	11.049	7.844	7.086	5.031	May
250	3.923	2.785	.553	11.558	8.206001	7.086	5.031	May
200	4.512	3.203	.636	12.033	8.543	7.086	5.031	May
150	5.124	3.638	.723	12.48	8.859999	7.086	5.031	May
100	5.758	4.088	.812	12.904	9.161	7.086	5.031	May
50	6.412	4.552	.904	13.305	9.446	7.086	5.031	May
0	7.086	5.031	1	13.691	9.72	7.086	5.031	May

Losses kWh/m2/d	Avail.Heat no clouds kWh/m2/d	Avail.Heat + clouds kWh/m2/d	Efficiency	Opera.Time no clouds hours/d	Opera.Time + clouds hours/d	Avail.Sun no clouds kWh/m2/d	Avail.Sun + clouds kWh/m2/d	
800	0	0	0	0	0	6.92	5.466	June
750	0	0	0	0	0	6.92	5.466	June
700	.015	.011	.002	1.771	1.399	6.92	5.466	June
650	.184	.145	.026	4.614	3.645	6.92	5.466	June
600	.455	.359	.065	6.156	4.863	6.92	5.466	June
550	.792	.625	.114	7.311	5.775	6.92	5.466	June
500	1.181	.932	.17	8.245	6.513	6.92	5.466	June
450	1.613	1.274	.233	9.05	7.149	6.92	5.466	June
400	2.083	1.645	.301	9.777001	7.723	6.92	5.466	June
350	2.588	2.044	.374	10.438	8.246	6.92	5.466	June
300	3.124	2.467	.451	11.053	8.731	6.92	5.466	June
250	3.691	2.915	.533	11.628	9.186	6.92	5.466	June
200	4.285	3.385	.619	12.173	9.616	6.92	5.466	June
150	4.906	3.875	.709	12.698	10.031	6.92	5.466	June
100	5.553	4.386	.802	13.2	10.428	6.92	5.466	June
50	6.224	4.916	.899	13.688	10.813	6.92	5.466	June
0	6.92	5.466	1	14.162	11.137	6.92	5.466	June

Losses kWh/m2/d	Avail.Heat no clouds kWh/m2/d	Avail.Heat + clouds kWh/m2/d	Efficiency	Opera.Time no clouds hours/d	Opera.Time + clouds hours/d	Avail.Sun no clouds kWh/m2/d	Avail.Sun + clouds kWh/m2/d	
800	0	0	0	0	0	7.055	6.067	July
750	0	0	0	0	0	7.055	6.067	July
700	.054	.046	.007	2.989	2.57	7.055	6.067	July
650	.262	.225	.037	5.127	4.409	7.055	6.067	July
600	.554	.476	.078	6.501	5.59	7.055	6.067	July
550	.905	.778	.128	7.568	6.508	7.055	6.067	July
500	1.306	1.123	.185	8.453	7.269	7.055	6.067	July
450	1.747	1.502	.247	9.225001	7.933	7.055	6.067	July
400	2.225	1.913	.315	9.915	8.526	7.055	6.067	July
350	2.736	2.352	.387	10.544	9.067001	7.055	6.067	July
300	3.277	2.818	.464	11.122	9.564	7.055	6.067	July
250	3.846	3.307	.545	11.66	10.027	7.055	6.067	July
200	4.441	3.819	.629	12.168	10.464	7.055	6.067	July
150	5.061	4.352	.717	12.647	10.876	7.055	6.067	July
100	5.704	4.905	.808	13.099	11.265	7.055	6.067	July
50	6.369	5.477	.902	13.536	11.64	7.055	6.067	July
0	7.055	6.067	1	13.952	11.998	7.055	6.067	July

Losses kWh/m2/d	Avail.Heat no clouds kWh/m2/d	Avail.Heat + clouds kWh/m2/d	Efficiency	Opera.Time no clouds hours/d	Opera.Time + clouds hours/d	Avail.Sun no clouds kWh/m2/d	Avail.Sun + clouds kWh/m2/d	
800	0	0	0	0	0	6.315	5.178	August
750	0	0	0	0	0	6.315	5.178	August
700	.001	0	0	.474	.388	6.315	5.178	August
650	.118	.096	.018	3.81	3.124	6.315	5.178	August
600	.352	.288	.055	5.461	4.478	6.315	5.178	August
550	.655	.537	.103	6.617	5.425	6.315	5.178	August
500	1.009	.827	.159	7.54	6.182	6.315	5.178	August
450	1.405	1.152	.222	8.334	6.833	6.315	5.178	August
400	1.839	1.507	.291	9.03	7.404	6.315	5.178	August
350	2.305	1.89	.364	9.653	7.915	6.315	5.178	August
300	2.802	2.297	.443	10.239	8.395001	6.315	5.178	August
250	3.326	2.727	.526	10.776	8.836001	6.315	5.178	August
200	3.877	3.179	.613	11.279	9.248	6.315	5.178	August
150	4.452	3.65	.705	11.758	9.641	6.315	5.178	August
100	5.051	4.141	.799	12.217	10.017	6.315	5.178	August
50	5.672	4.651	.898	12.654	10.376	6.315	5.178	August
0	6.315	5.178	1	13.076	10.722	6.315	5.178	August

Losses kWh/m2/d	Avail.Heat no clouds kWh/m2/d	Avail.Heat + clouds kWh/m2/d	Efficiency	Opera.Time no clouds hours/d	Opera.Time + clouds hours/d	Avail.Sun no clouds kWh/m2/d	Avail.Sun + clouds kWh/m2/d	
800	0	0	0	0	0	6.118	4.649	September
750	0	0	0	0	0	6.118	4.649	September
700	.028	.021	.004	1.858	1.412	6.118	4.649	September
650	.188	.142	.03	4.244	3.225	6.118	4.649	September
600	.435	.33	.07	5.596	4.252	6.118	4.649	September
550	.741	.563	.12	6.604	5.019	6.118	4.649	September
500	1.091	.829	.177	7.423	5.641	6.118	4.649	September
450	1.479	1.124	.241	8.121	6.171	6.118	4.649	September
400	1.9	1.444	.31	8.731	6.635	6.118	4.649	September
350	2.35	1.785	.383	9.276	7.049	6.118	4.649	September
300	2.825	2.147	.461	9.770001	7.425	6.118	4.649	September
250	3.324	2.526	.543	10.224	7.77	6.118	4.649	September
200	3.845	2.922	.628	10.641	8.087	6.118	4.649	September
150	4.386	3.333	.716	11.03	8.382	6.118	4.649	September
100	4.946	3.758	.808	11.338	8.654	6.118	4.649	September
50	5.523	4.197	.902	11.735	8.918	6.118	4.649	September
0	6.118	4.649	1	12.06	9.165	6.118	4.649	September

Losses	Avail.Heat no clouds	Avail.Heat + clouds	Efficiency	Opera.Time no clouds	Opera.Time + clouds	Avail.Sun no clouds	Avail.Sun + clouds	
kWh/m2/d	kWh/m2/d	kWh/m2/d		hours/d	hours/d	kWh/m2/d	kWh/m2/d	
800	0	0	0	0	0	5.694	3.53	October
750	0	0	0	0	0	5.694	3.53	October
700	.033	.02	.005	1.723	1.068	5.694	3.53	October
650	.18	.111	.03	3.97	2.461	5.694	3.53	October
600	.412	.255	.071	5.275	3.27	5.694	3.53	October
550	.7	.434	.122	6.241	3.869	5.694	3.53	October
500	1.032	.639	.18	7.012	4.347	5.694	3.53	October
450	1.398	.866	.244	7.652	4.744	5.694	3.53	October
400	1.794	1.112	.314	8.202	5.085	5.694	3.53	October
350	2.215	1.373	.388	8.681	5.382	5.694	3.53	October
300	2.659	1.648	.466	9.109	5.647	5.694	3.53	October
250	3.124	1.936	.547	9.495	5.886	5.694	3.53	October
200	3.607	2.236	.632	9.844	6.103	5.694	3.53	October
150	4.106	2.545	.72	10.168	6.304	5.694	3.53	October
100	4.621	2.865	.811	10.465	6.488	5.694	3.53	October
50	5.151	3.193	.904	10.75	6.665	5.694	3.53	October
0	5.694	3.53	1	11.006	6.823	5.694	3.53	October

Losses	Avail.Heat no clouds	Avail.Heat + clouds	Efficiency	Opera.Time no clouds	Opera.Time + clouds	Avail.Sun no clouds	Avail.Sun + clouds	
kWh/m2/d	kWh/m2/d	kWh/m2/d		hours/d	hours/d	kWh/m2/d	kWh/m2/d	
800	0	0	0	0	0	4.906	2.943	November
750	0	0	0	0	0	4.906	2.943	November
700	0	0	0	.117	.07	4.906	2.943	November
650	.049	.029	$8.999999E-03$	2.179	1.307	4.906	2.943	November
600	.21	.126	.042	4.061	2.436	4.906	2.943	November
550	.442	.265	.089	5.183	3.109	4.906	2.943	November
500	.722	.433	.146	6.022	3.613	4.906	2.943	November
450	1.04	.624	.211	6.697	4.018	4.906	2.943	November
400	1.389	.833	.282	7.261	4.356	4.906	2.943	November
350	1.763	1.057	.356	7.741	4.644	4.906	2.943	November
300	2.161	1.296	.439	8.17	4.902	4.906	2.943	November
250	2.578	1.546	.524	8.548001	5.128	4.906	2.943	November
200	3.013	1.807	.613	8.895001	5.337	4.906	2.943	November
150	3.465	2.079	.705	9.206001	5.523	4.906	2.943	November
100	3.932	2.359	.801	9.494001	5.696	4.906	2.943	November
50	4.413	2.647	.899	9.764	5.858	4.906	2.943	November
0	4.906	2.943	1	10.016	6.009	4.906	2.943	November

Losses	Avail.Heat no clouds	Avail.Heat + clouds	Efficiency	Opera.Time no clouds	Opera.Time + clouds	Avail.Sun no clouds	Avail.Sun + clouds	
kWh/m2/d	kWh/m2/d	kWh/m2/d		hours/d	hours/d	kWh/m2/d	kWh/m2/d	
800	0	0	0	0	0	4.49	2.918	December
750	0	0	0	0	0	4.49	2.918	December
700	0	0	0	0	0	4.49	2.918	December
650	0	0	0	.252	.163	4.49	2.918	December
600	.102	.066	.022	3.184	2.069	4.49	2.918	December
550	.296	.192	.065	4.505	2.928	4.49	2.918	December
500	.545	.354	.121	5.449	3.541	4.49	2.918	December
450	.835	.542	.186	6.177	4.015	4.49	2.918	December
400	1.159	.753	.258	6.77	4.4	4.49	2.918	December
350	1.509	.98	.336	7.274	4.728	4.49	2.918	December
300	1.883	1.223	.419	7.708	5.01	4.49	2.918	December
250	2.278	1.48	.5070001	8.09	5.258	4.49	2.918	December
200	2.69	1.748	.599	8.432999	5.481	4.49	2.918	December
150	3.119	2.027	.674	8.747	5.685	4.49	2.918	December
100	3.563	2.315	.793	9.026	5.866	4.49	2.918	December
50	4.02	2.613	.895	9.288999	6.037	4.49	2.918	December
0	4.49	2.918	1	9.535	6.197	4.49	2.918	December

Losses kWh/m2/d	Avail.Heat no clouds kWh/m2/d	Avail.Heat + clouds kWh/m2/d	Efficiency	Opera.Time no clouds hours/d	Opera.Time + clouds hours/d	Avail.Sun no clouds kWh/m2/d	Avail.Sun + clouds kWh/m2/d	
800	0	0	0	0	0	5.923	4.075	Yearly
750	0	0	0	.061	.032	5.923	4.075	Yearly
700	.034	.022	.005	1.621	1.107	5.923	4.075	Yearly
650	.165	.113	.025	3.53	2.447	5.923	4.075	Yearly
600	.385	.265	.061	5.184	3.569	5.923	4.075	Yearly
550	.673	.462	.109	6.293	4.326	5.923	4.075	Yearly
500	1.009	.693	.166	7.156	4.917	5.923	4.075	Yearly
450	1.384	.952	.229	7.876	5.412	5.923	4.075	Yearly
400	1.793	1.232	.298	8.498999	5.84	5.923	4.075	Yearly
350	2.231	1.534	.373	9.048001	6.22	5.923	4.075	Yearly
300	2.695	1.853	.452	9.547	6.565	5.923	4.075	Yearly
250	3.183	2.189	.534	10.001	6.879	5.923	4.075	Yearly
200	3.693	2.539	.621	10.42	7.17	5.923	4.075	Yearly
150	4.223	2.904	.711	10.812	7.442	5.923	4.075	Yearly
100	4.772	3.282	.804	11.177	7.697	5.923	4.075	Yearly
50	5.339	3.673	.9	11.524	7.939	5.923	4.075	Yearly
0	5.923	4.075	1	11.853	8.169	5.923	4.075	Yearly

Laufzeit= 2586.801   Ende: 16:03:48

Available Process Heat

for

Clear Sky and Cloudy Sky

Monthly Averages

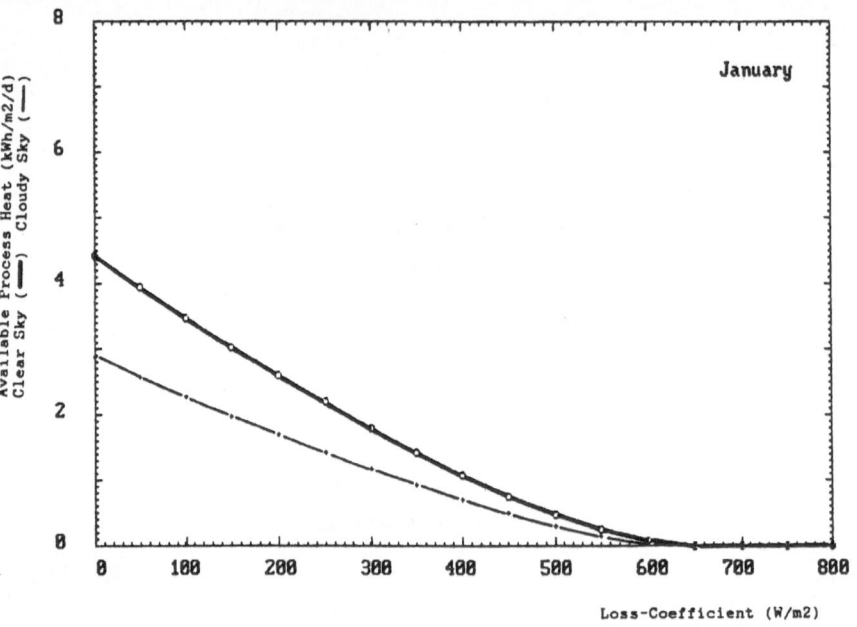

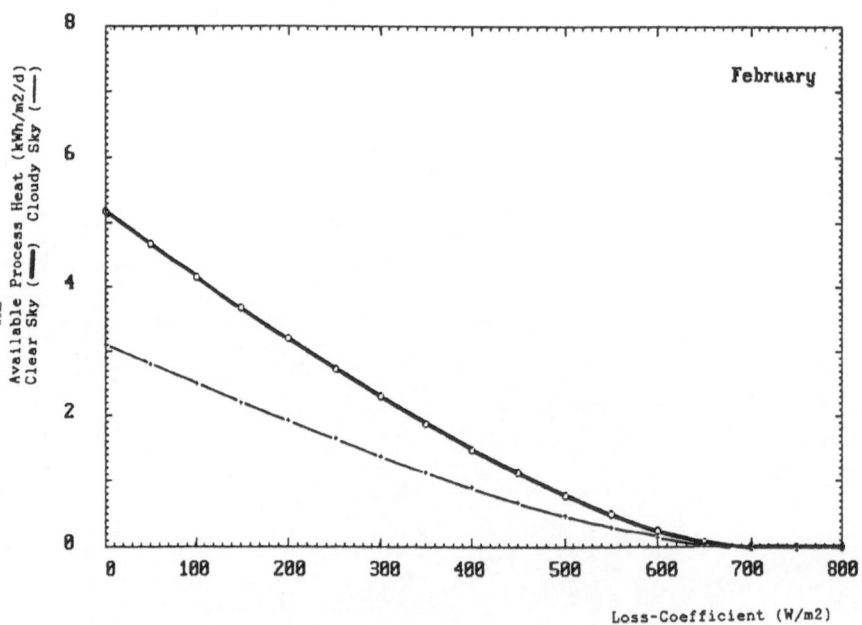

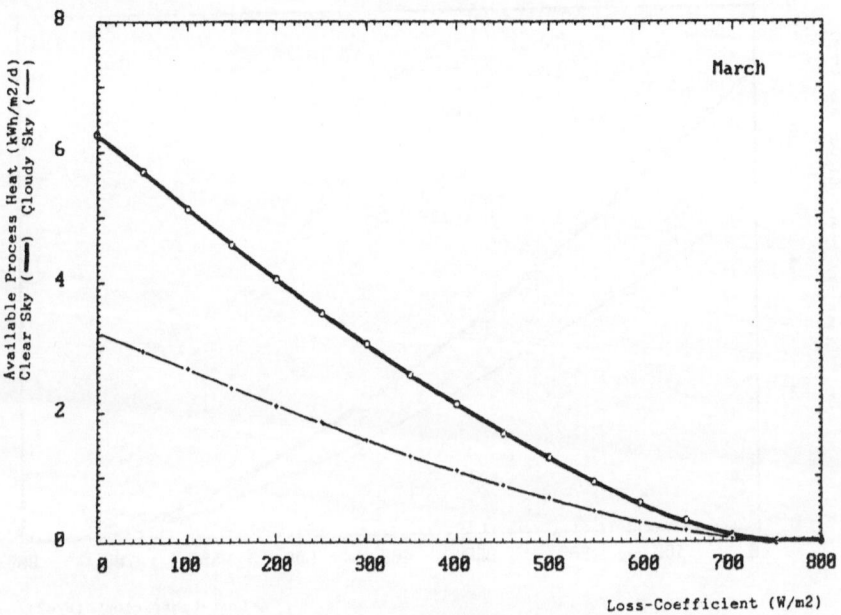

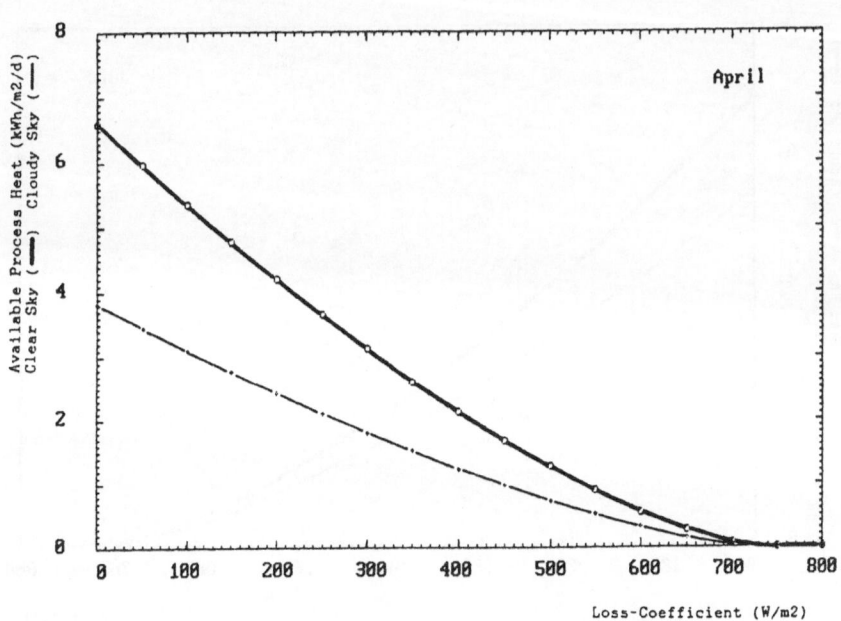

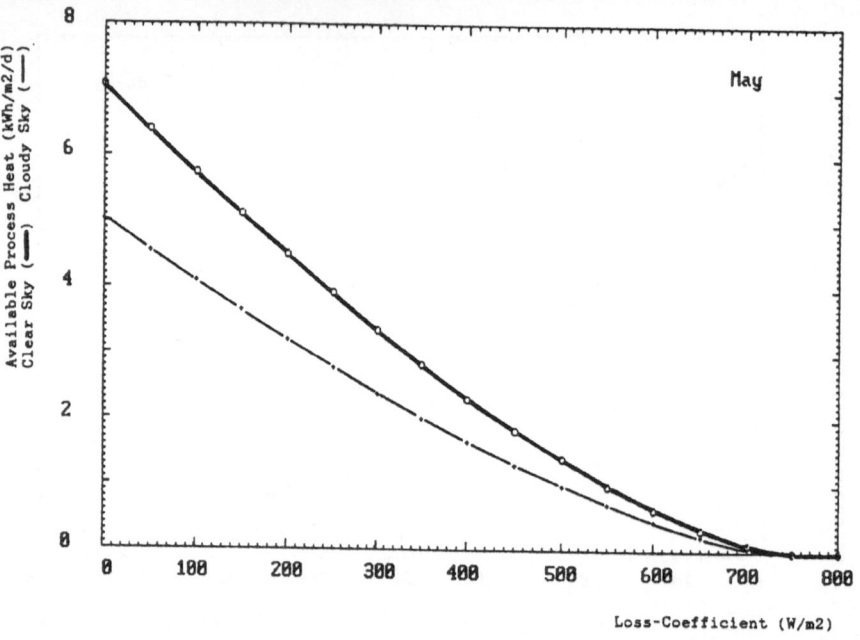

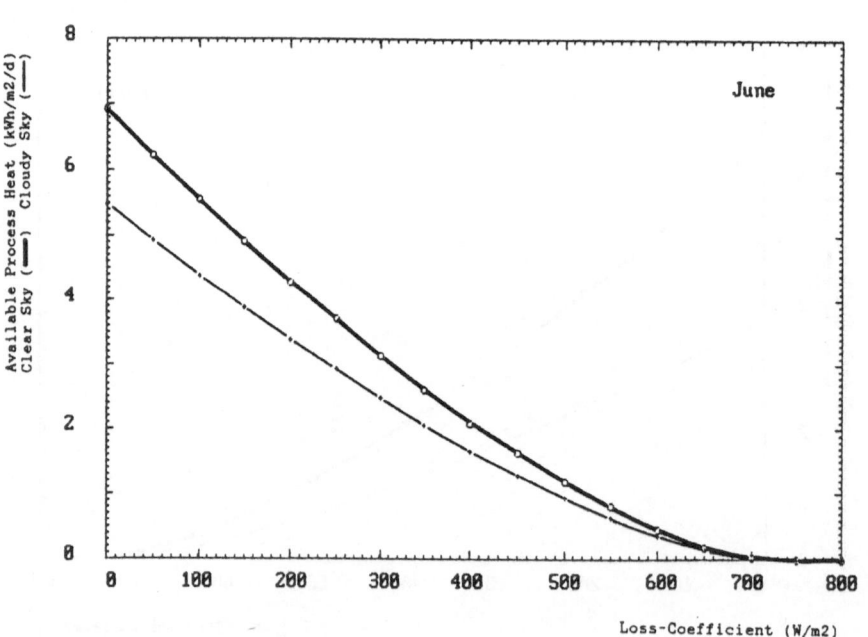

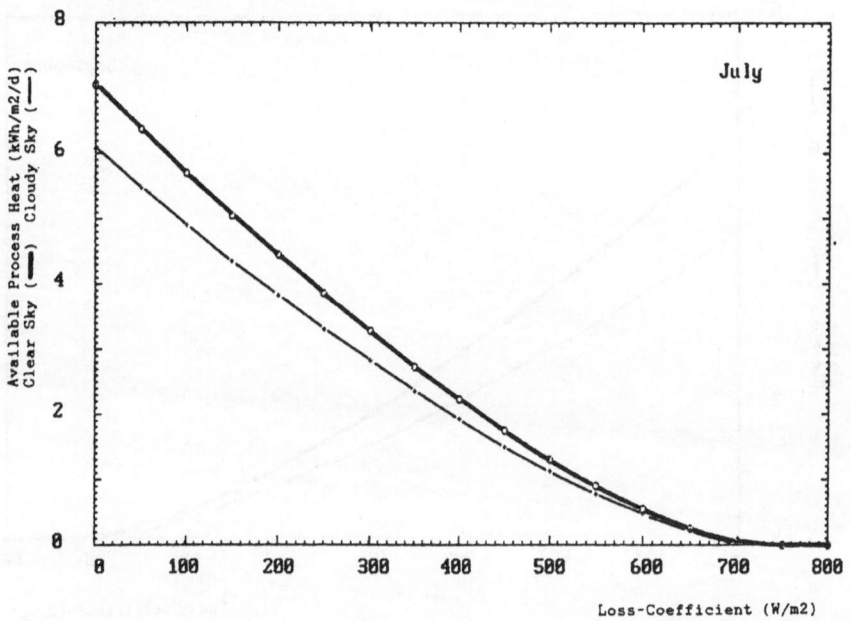

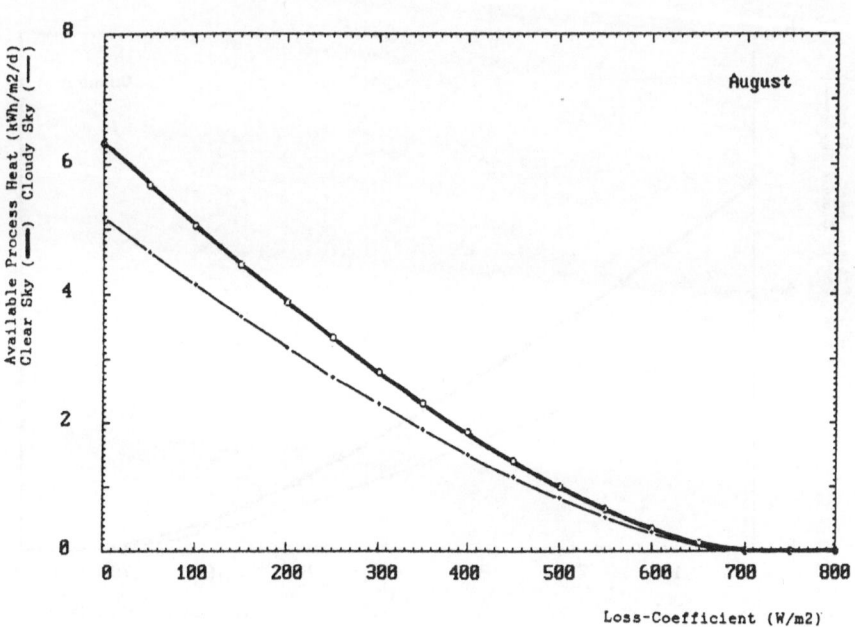

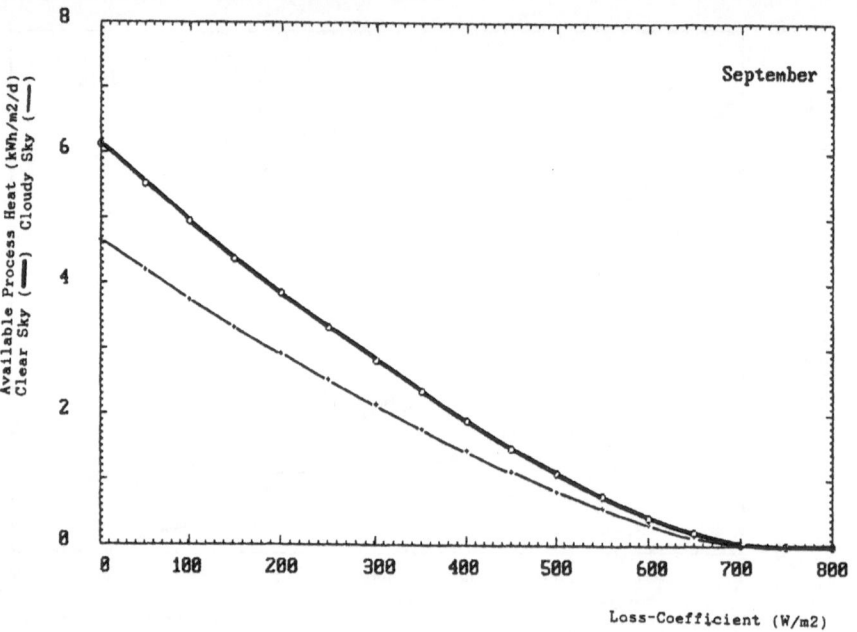

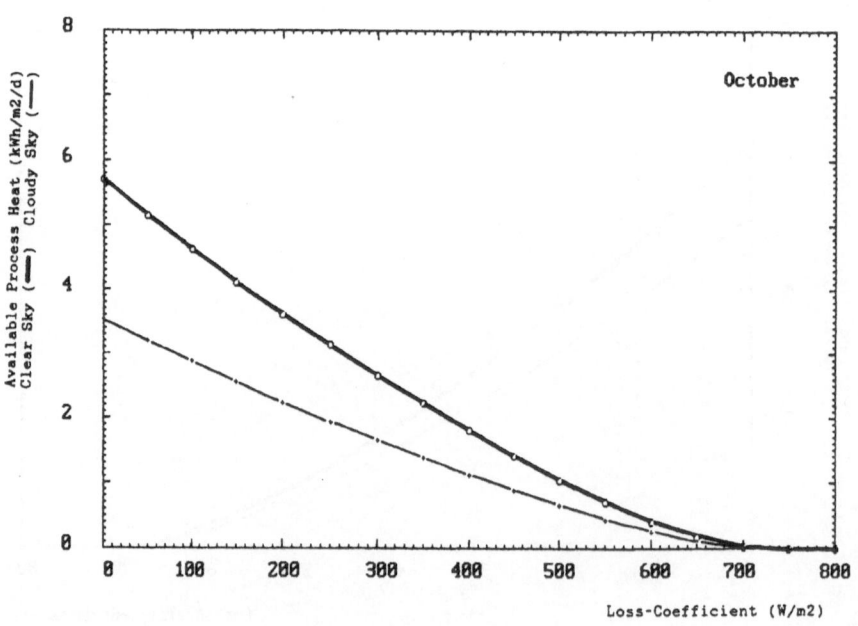

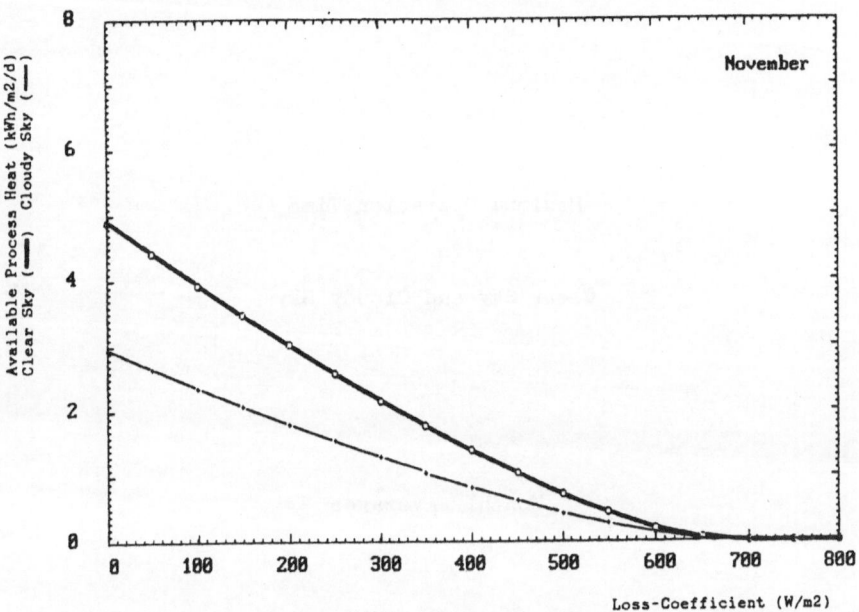

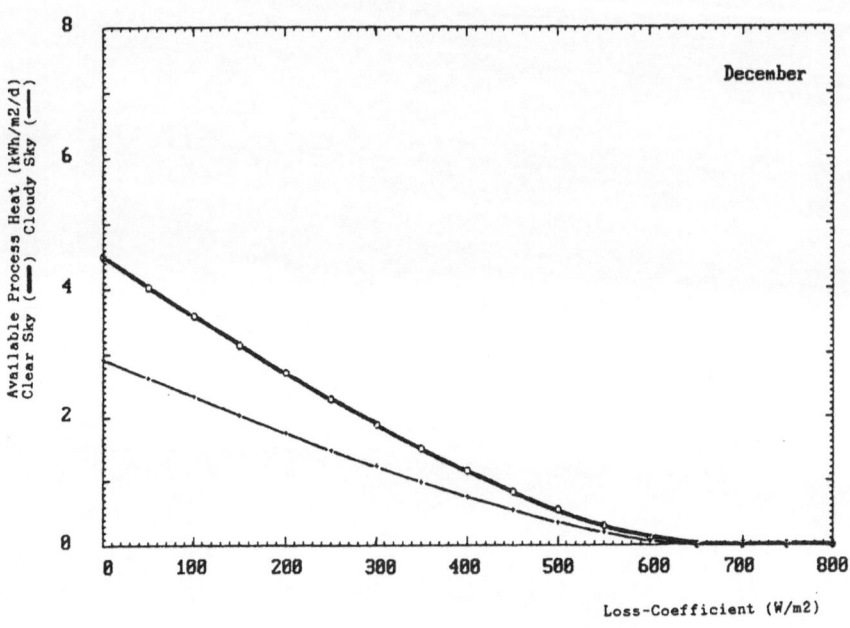

Maximum Operation Time

for

Clear Sky and Cloudy Sky

Monthly Averages

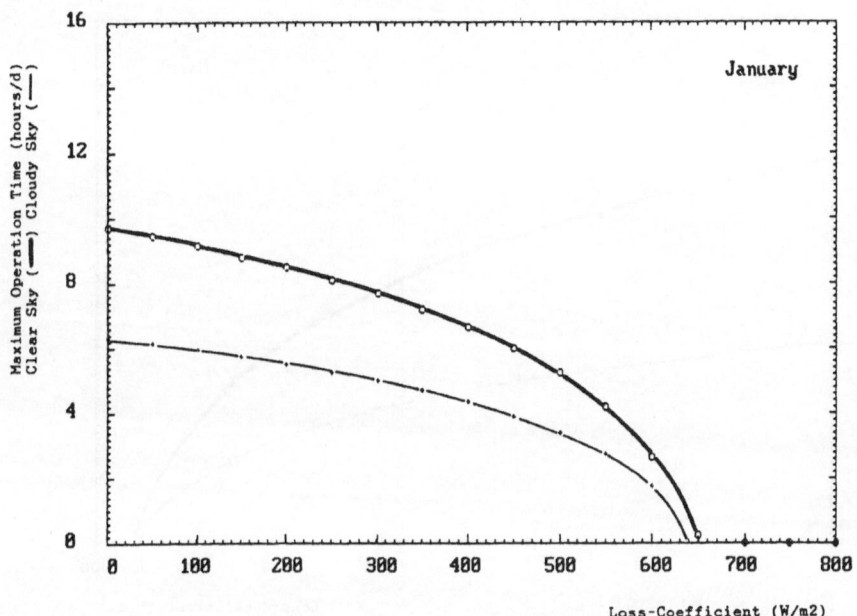

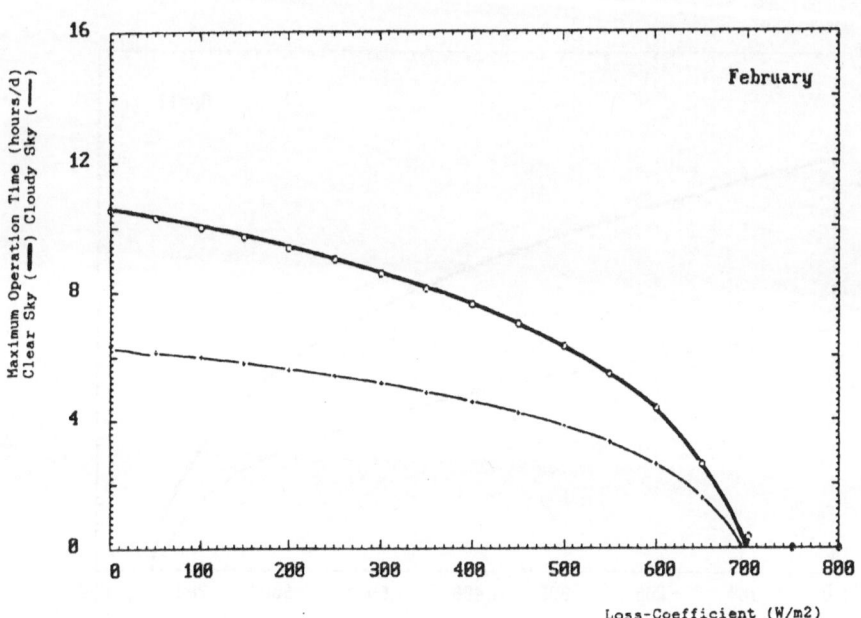

- 61 -

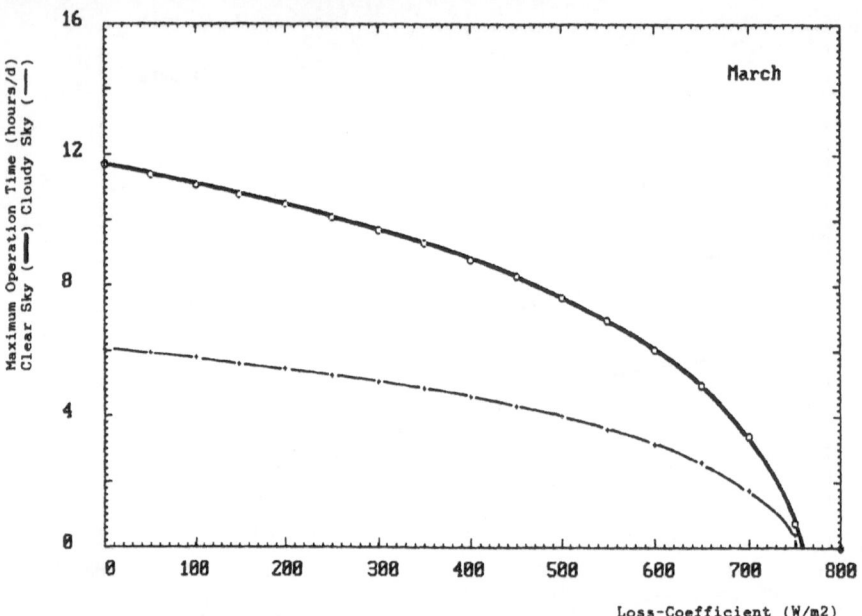

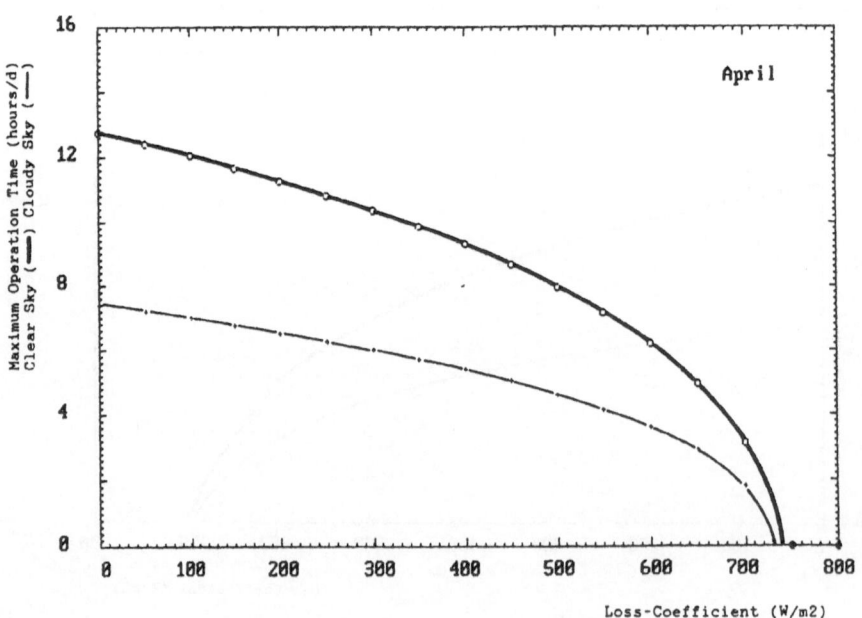

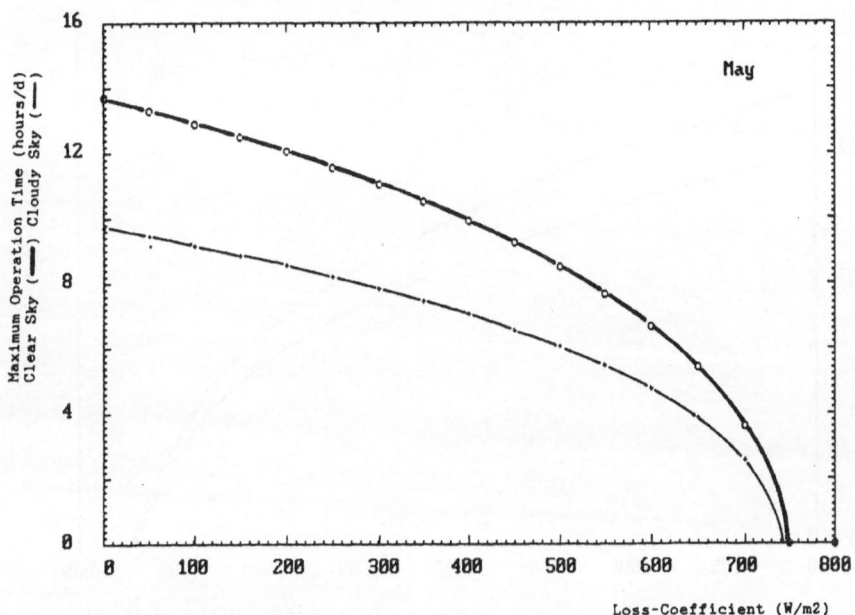

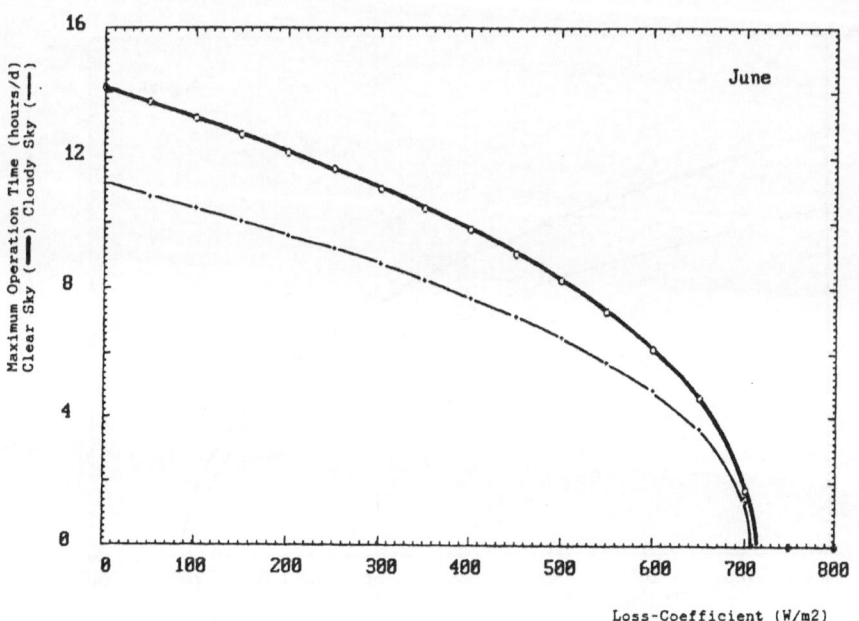

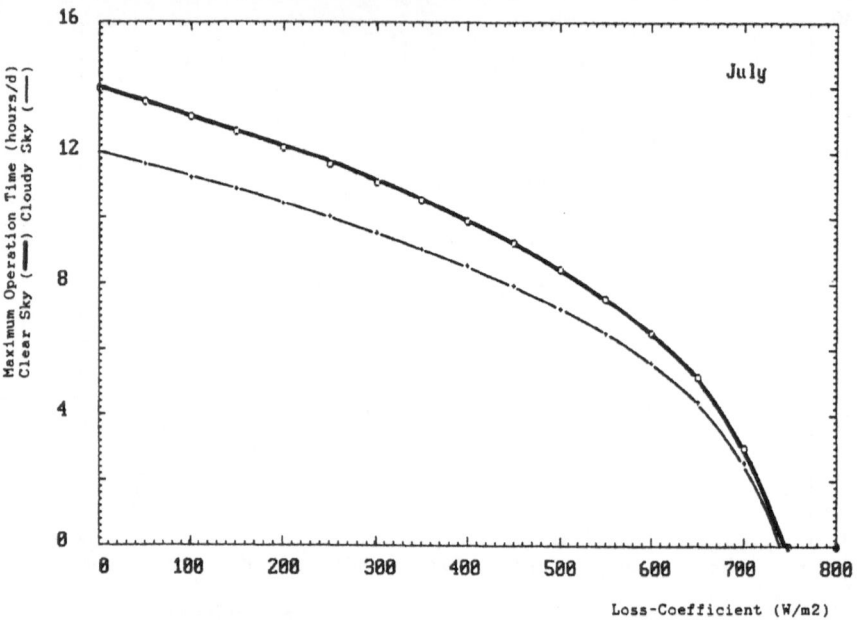

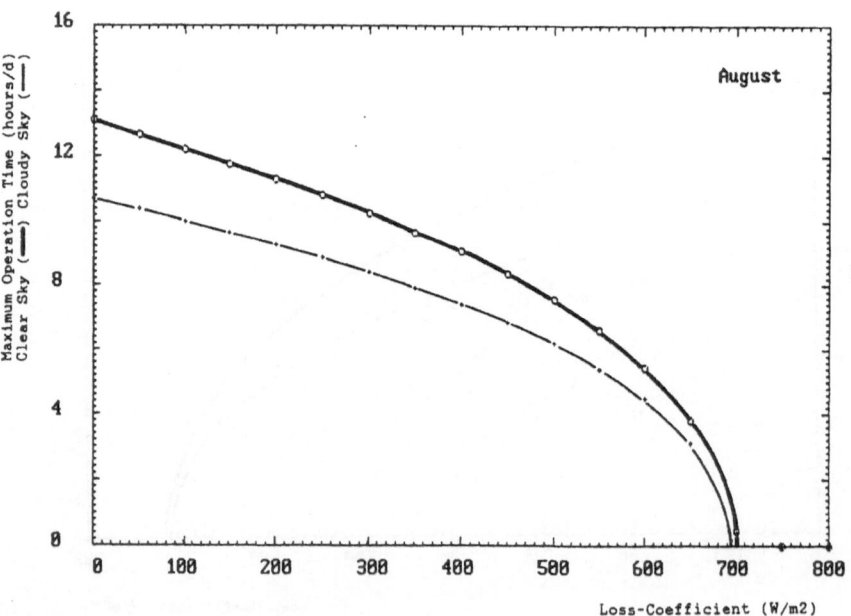

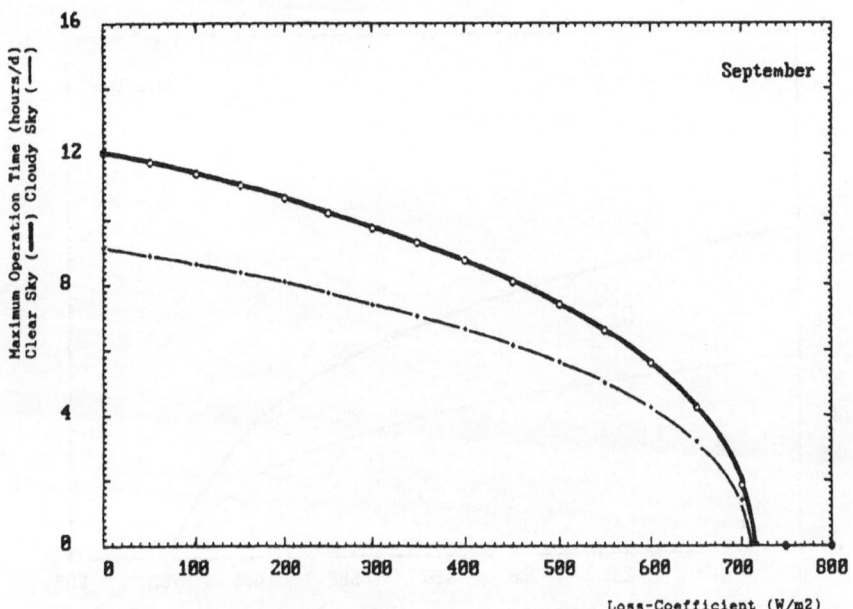

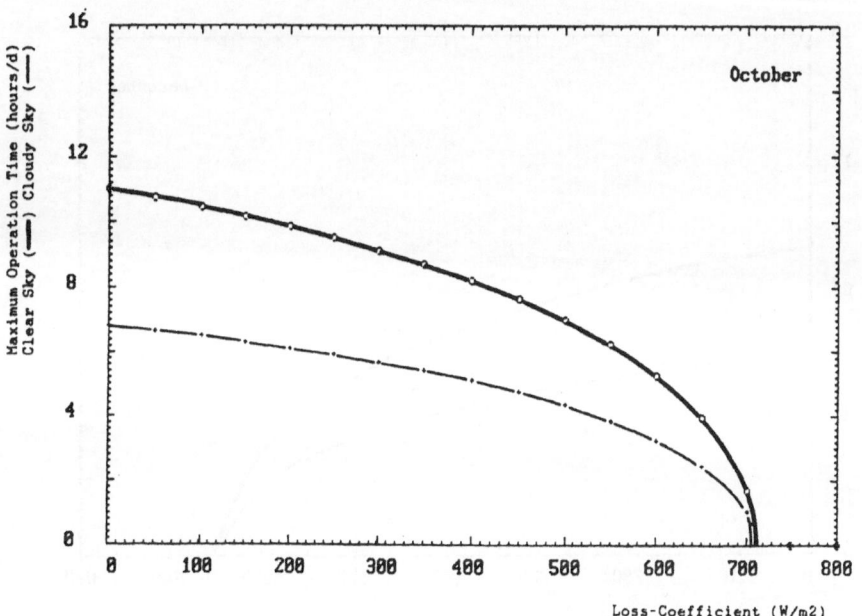

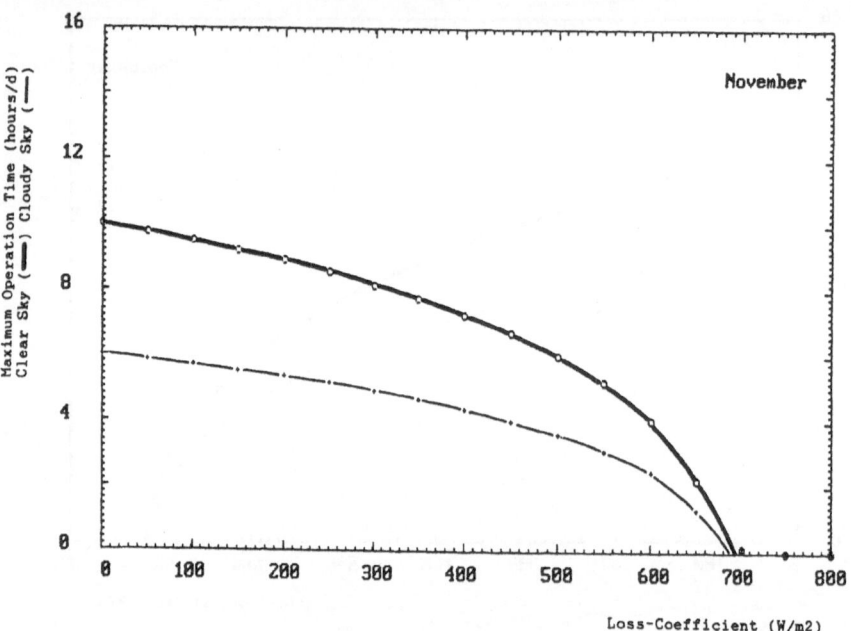

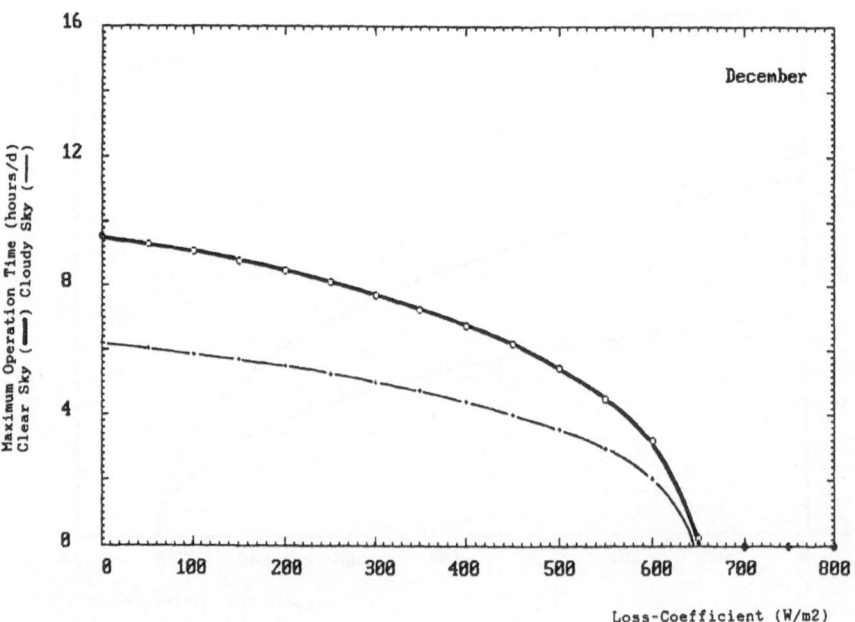

Yields

of

Process Heat Obtained/Solar Radiative Input

Monthly Averages

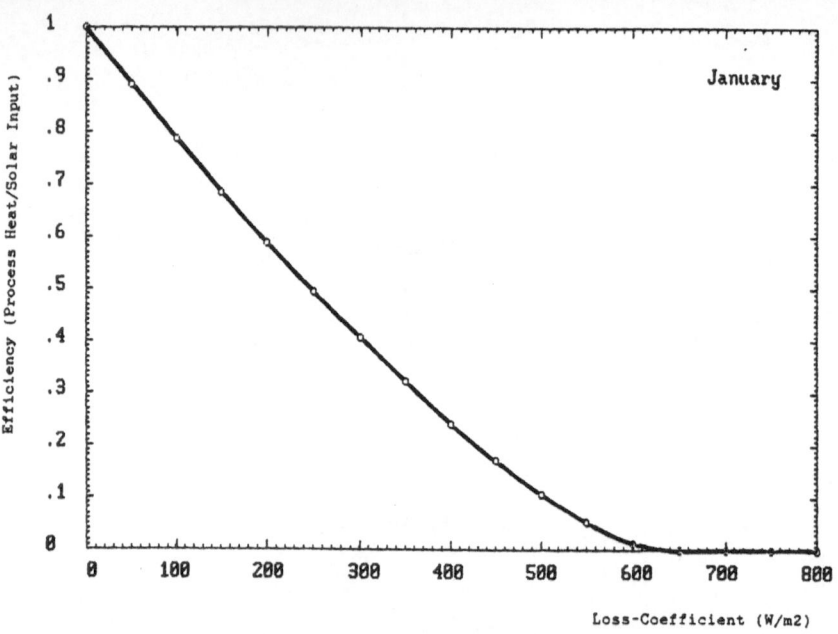

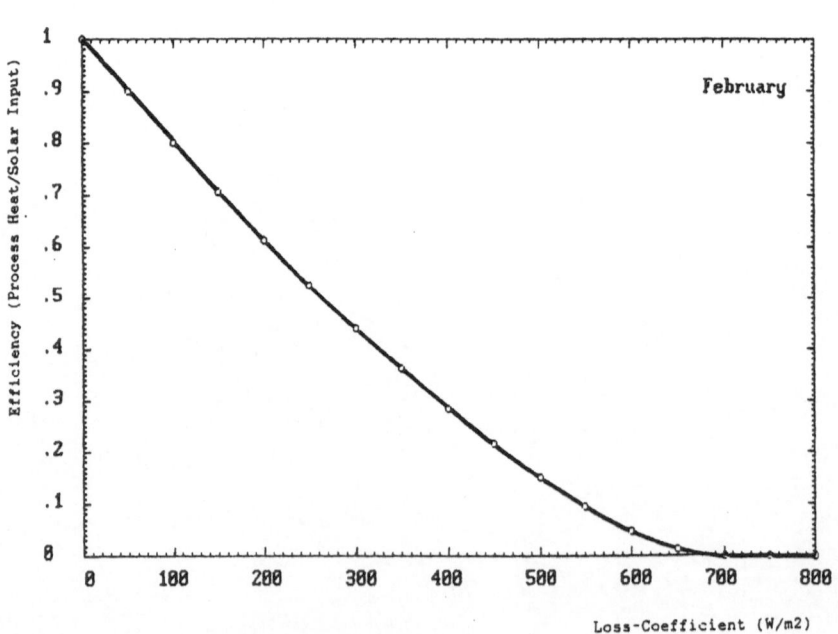

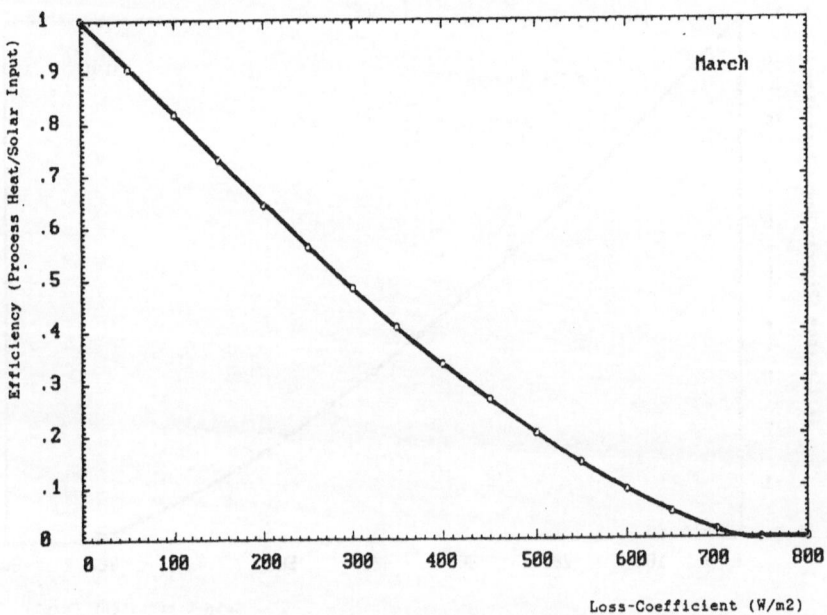

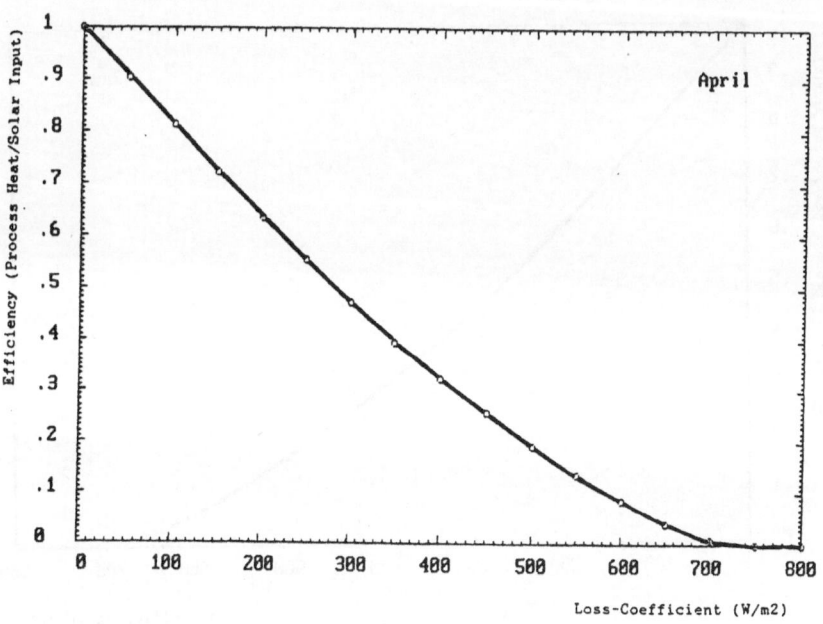

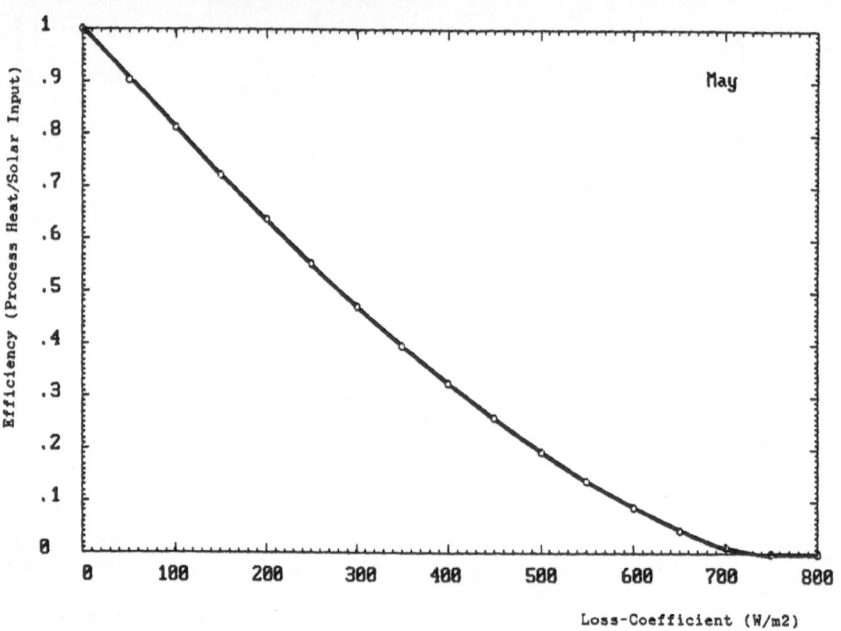

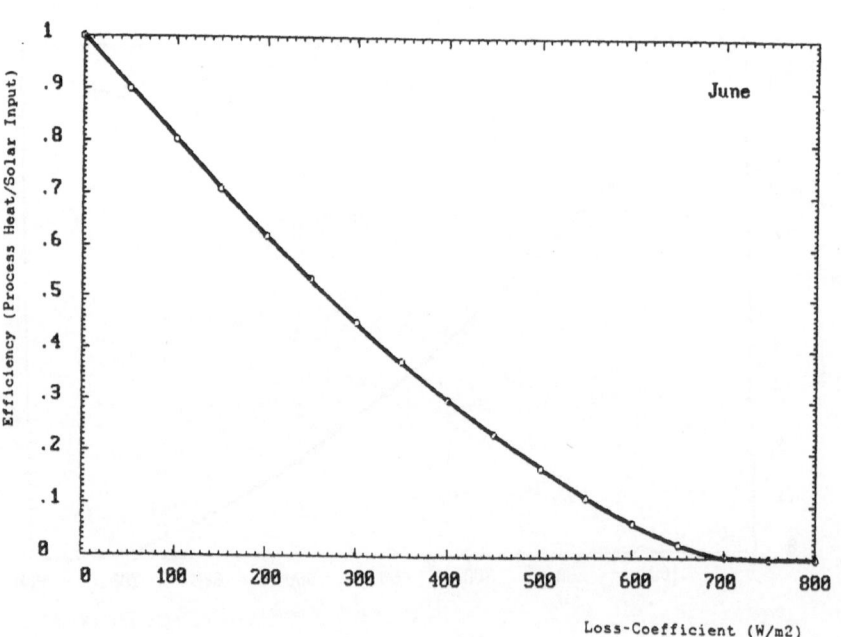

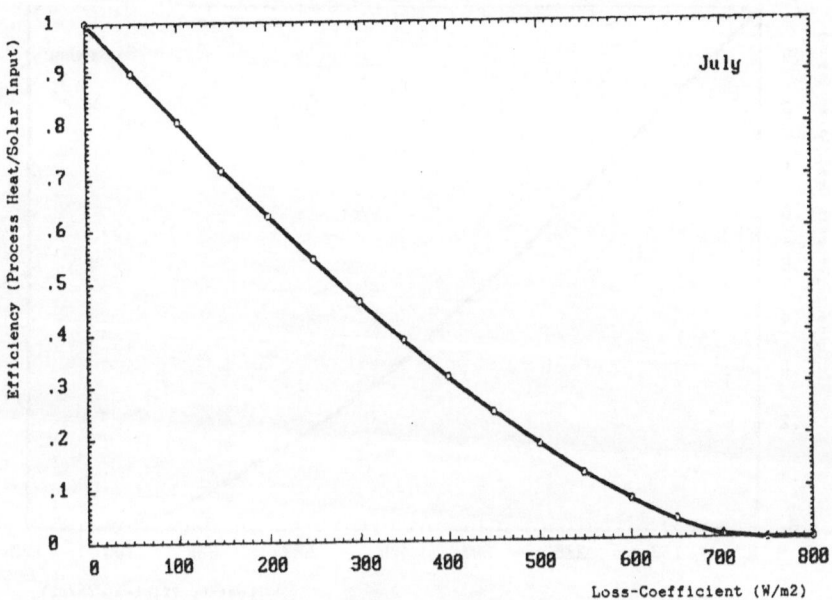

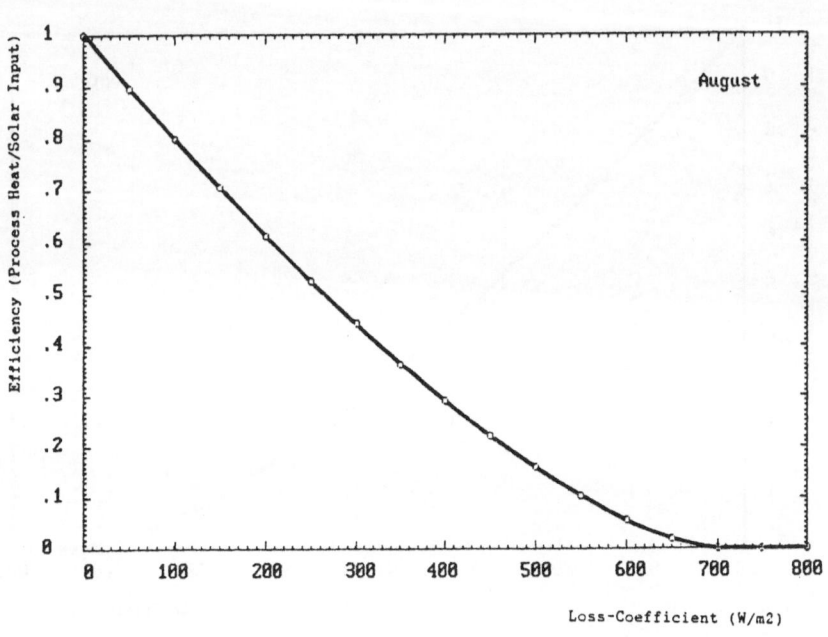

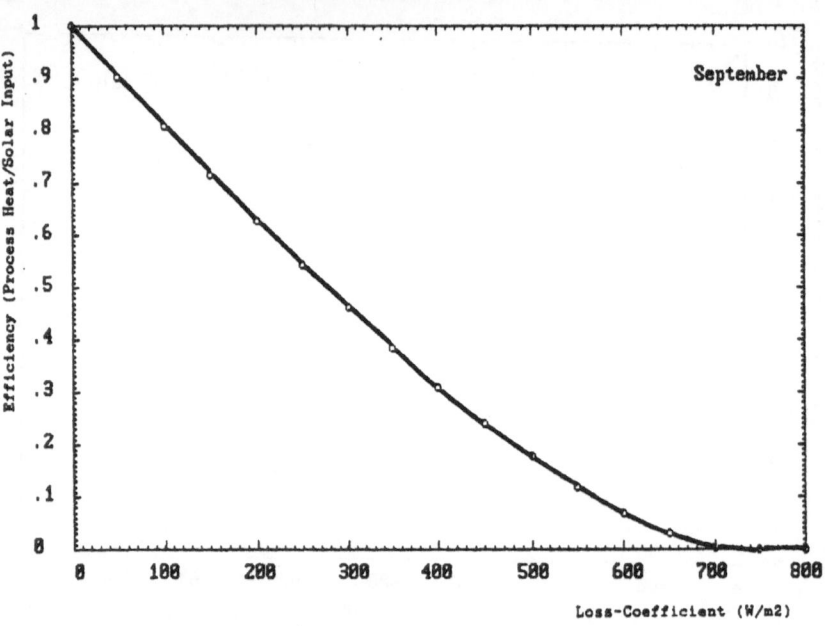

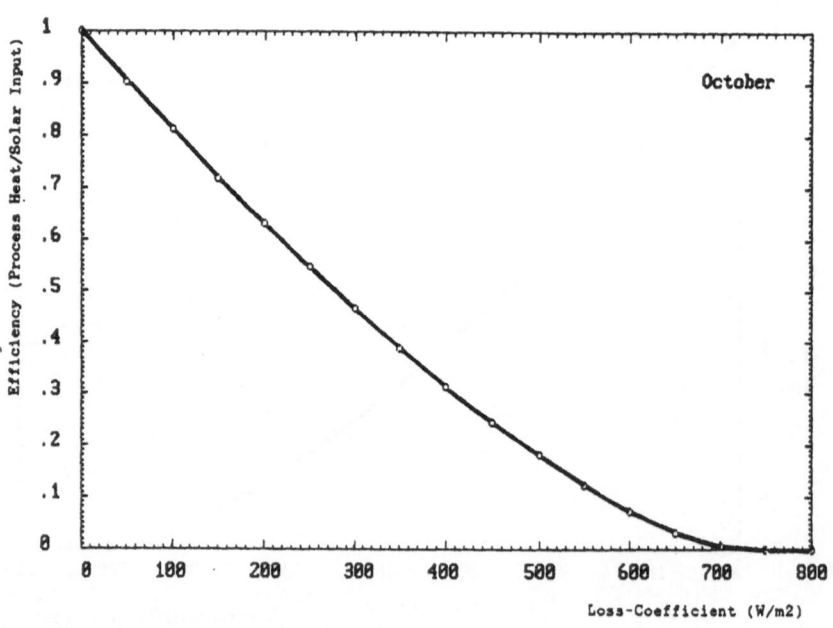

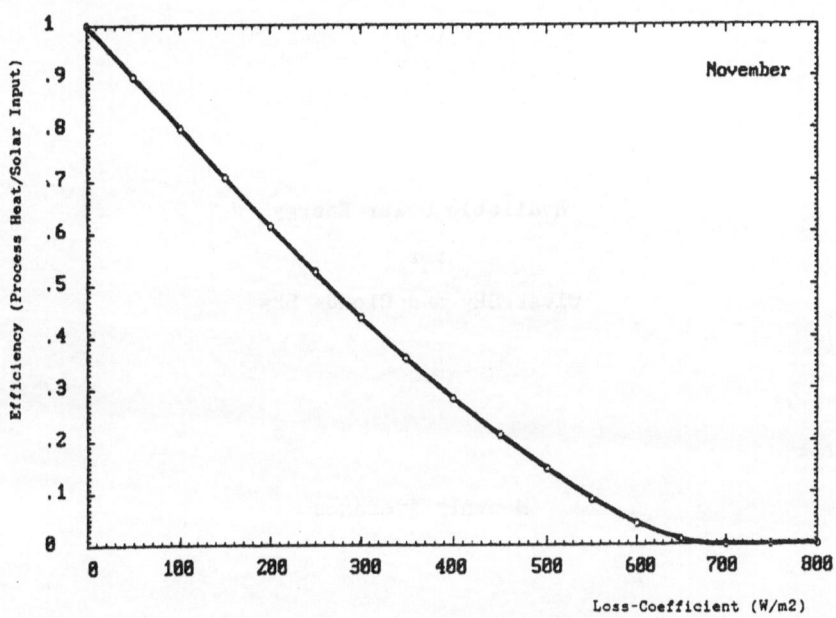

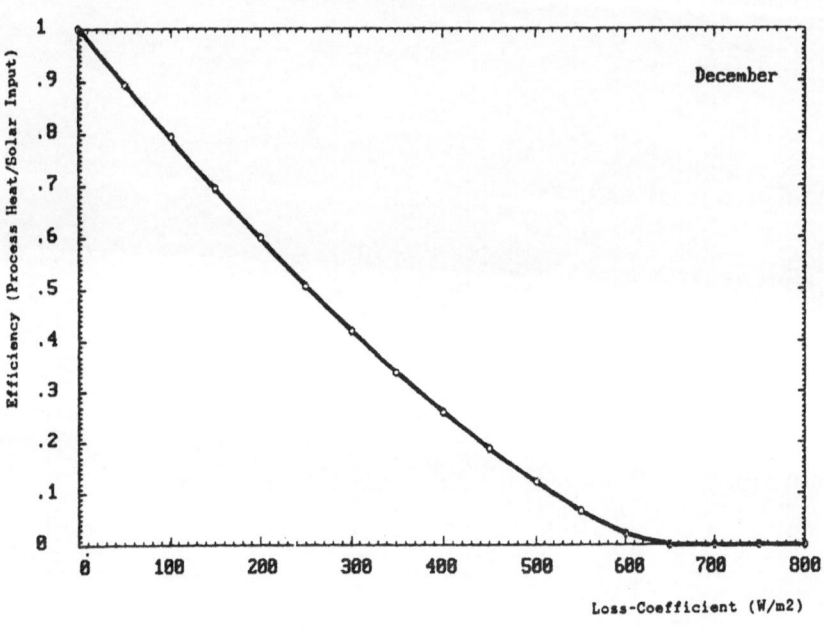

Available Solar Energy

for

Clear Sky and Cloudy Sky

Monthly Averages

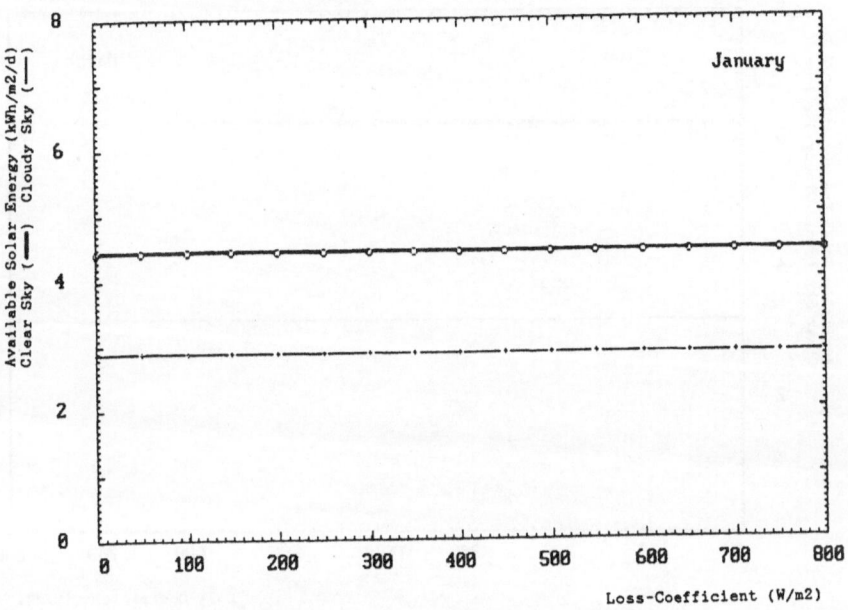

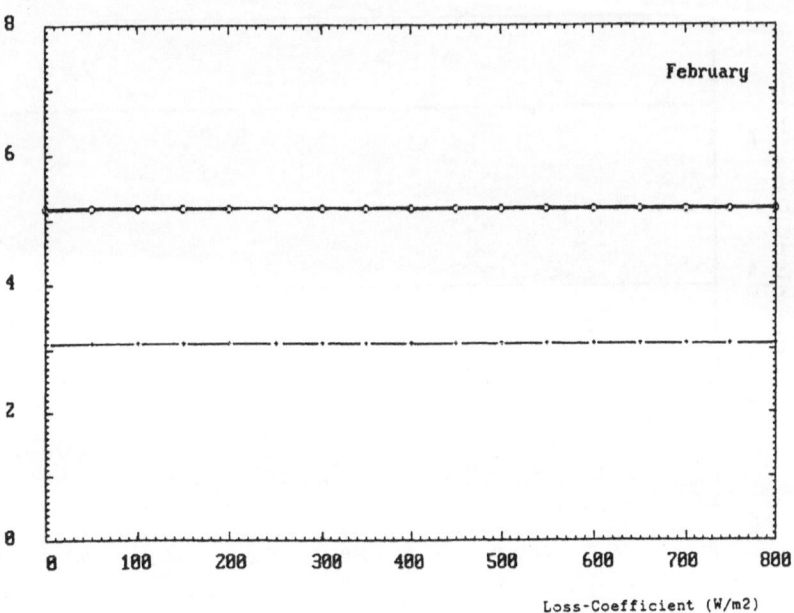

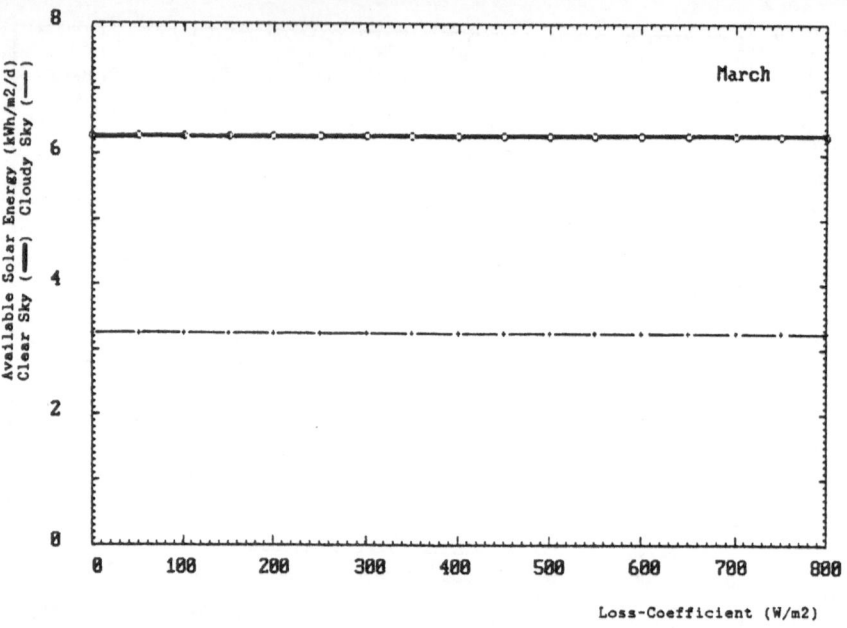

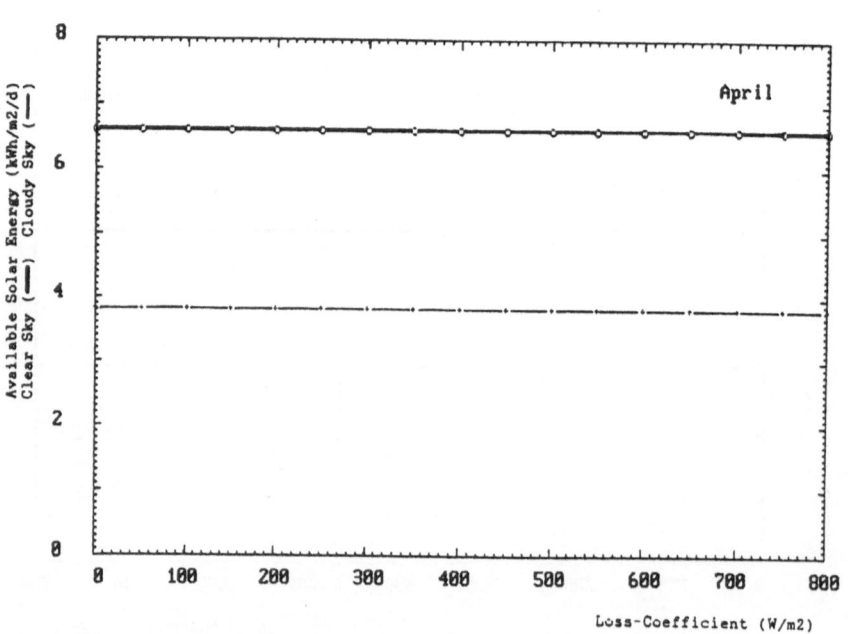

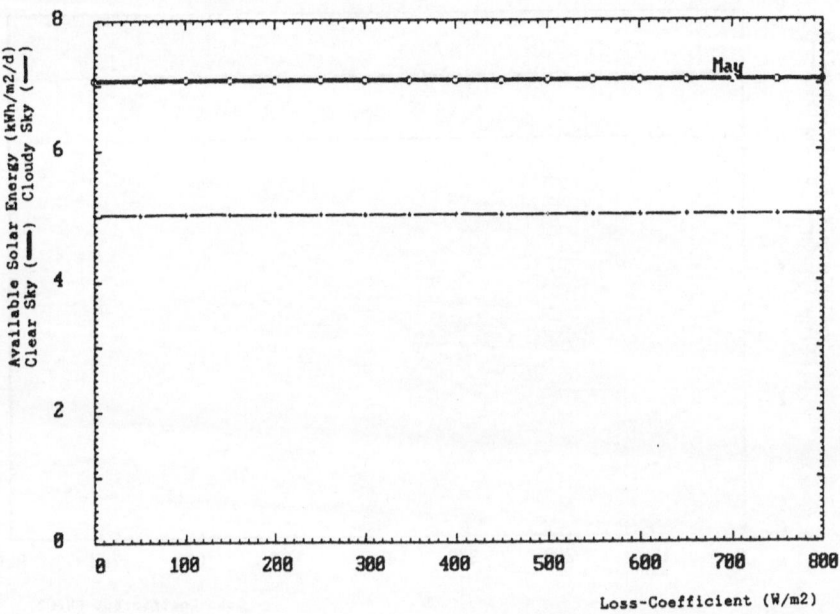

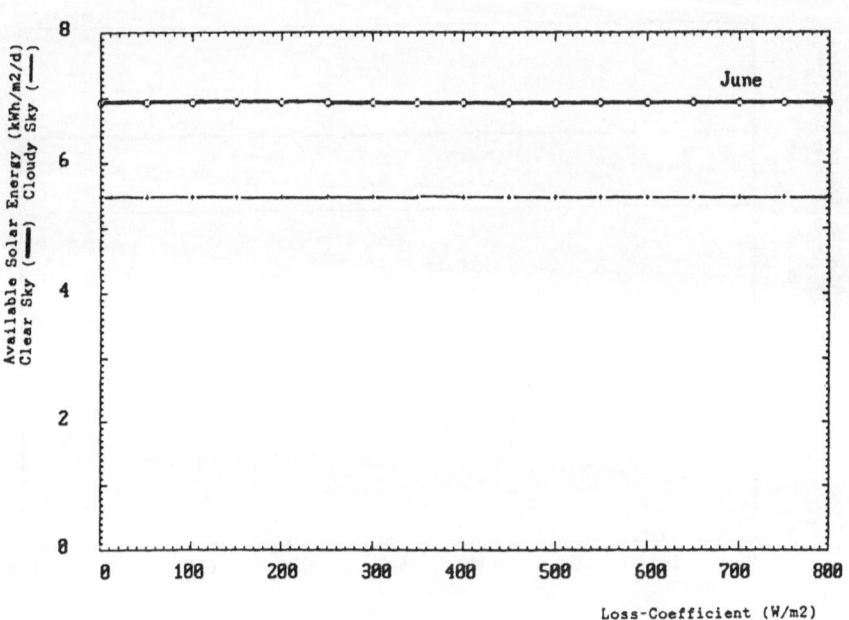

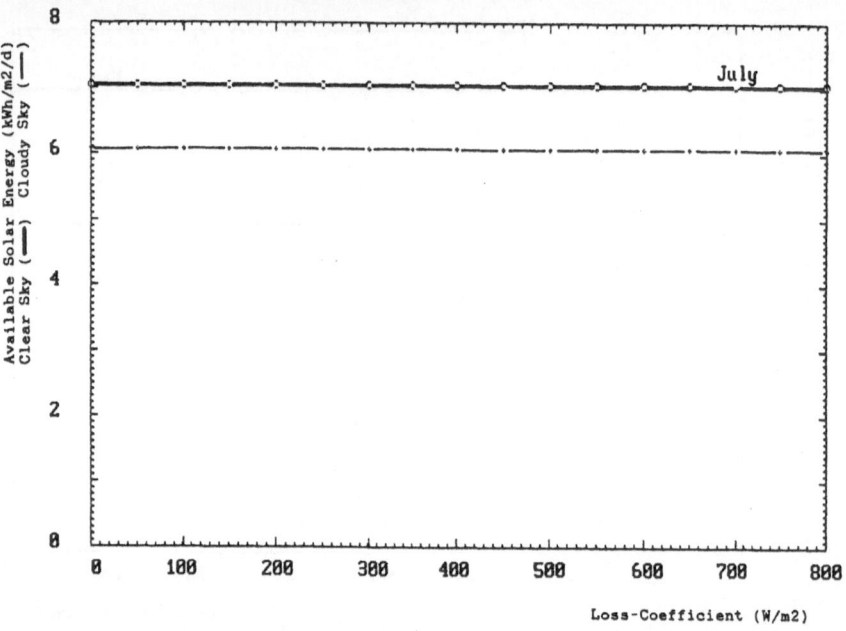

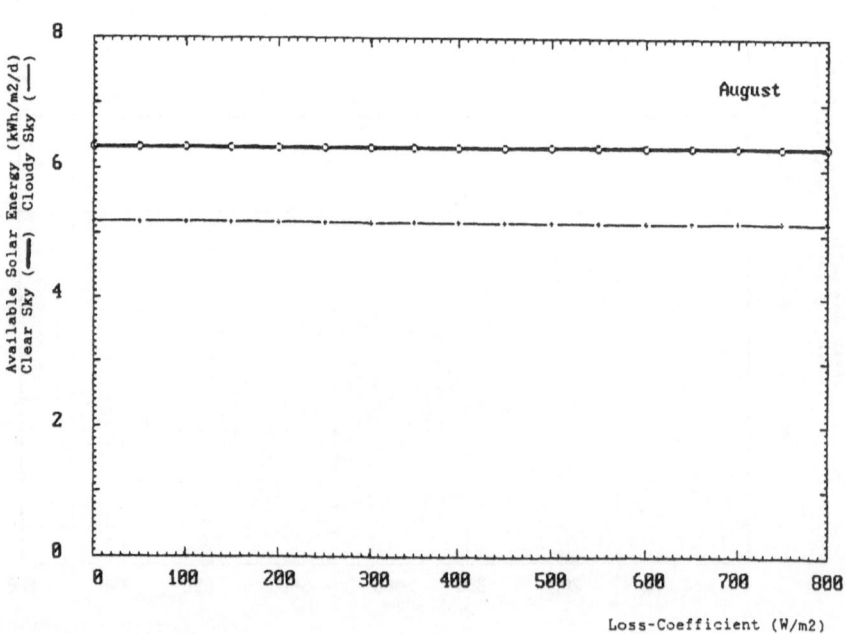

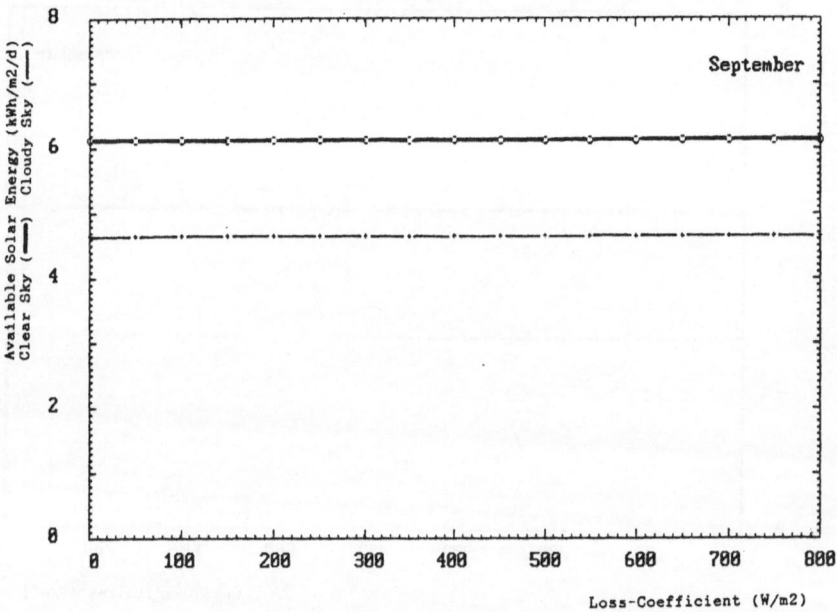

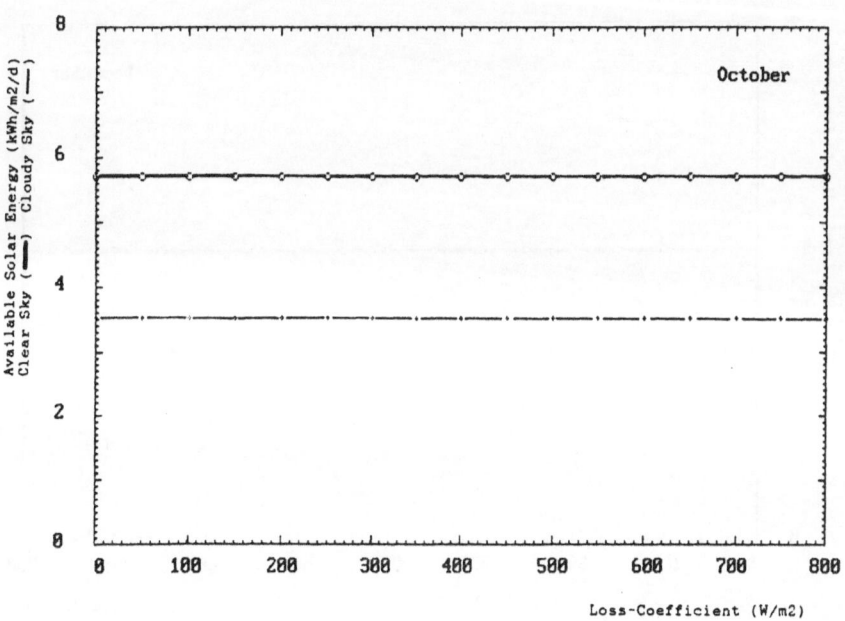

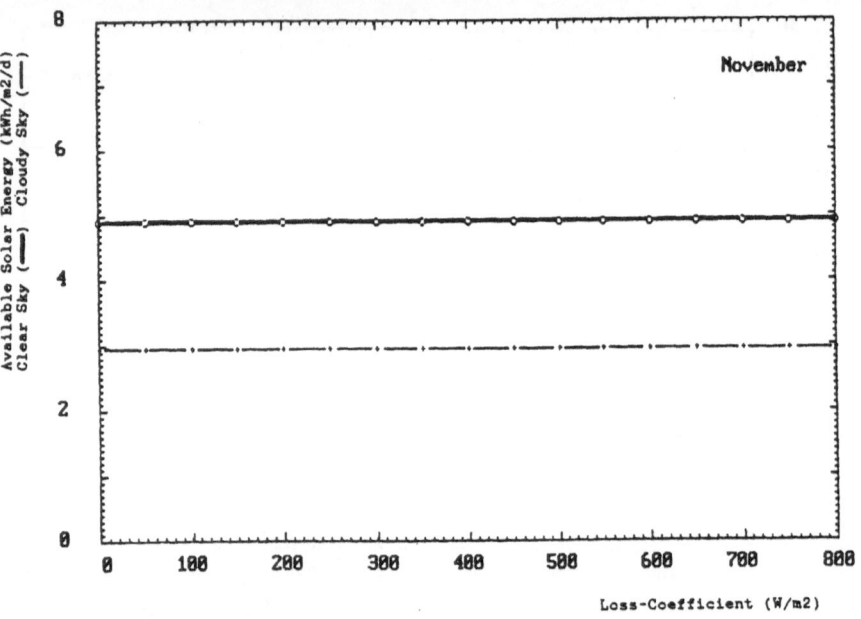

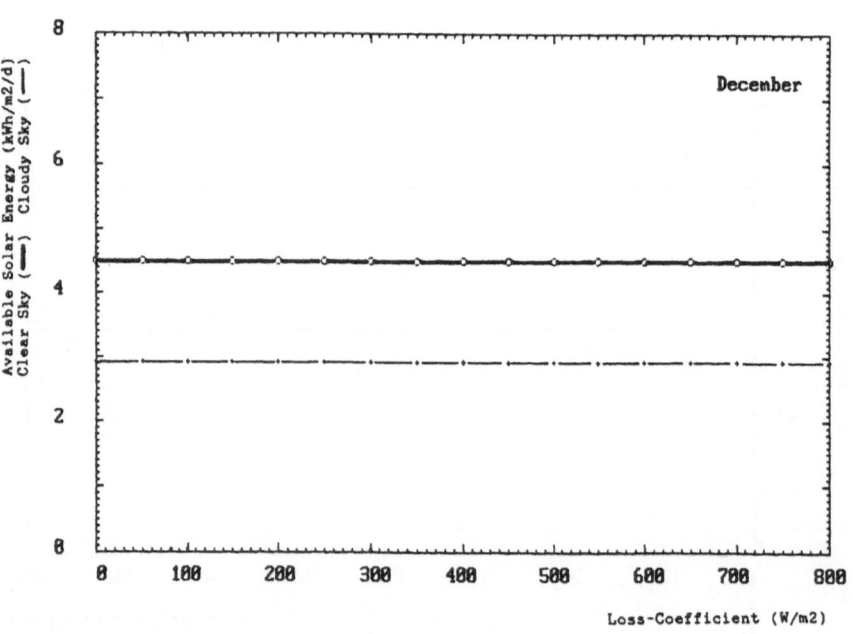

Available Process Heat

Maximum Operation Time

Yield of Process Heat from Solar Radiative Input

Available Solar Energy

for

Clear Sky and Cloudy Sky

Yearly Averages

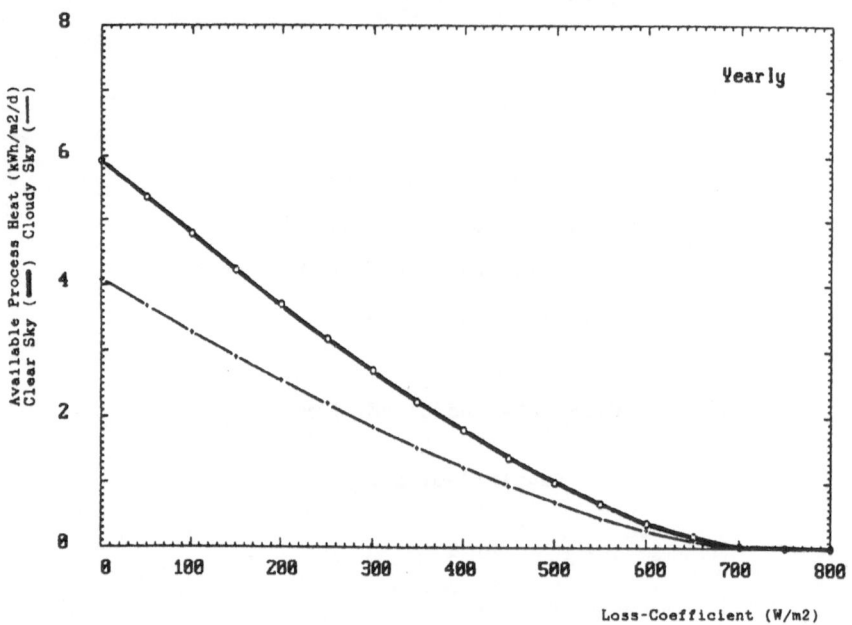

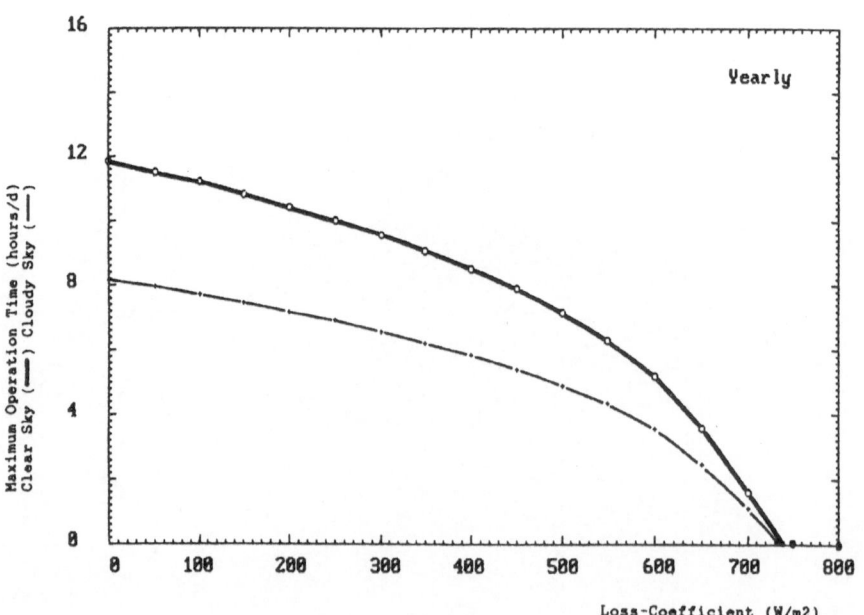

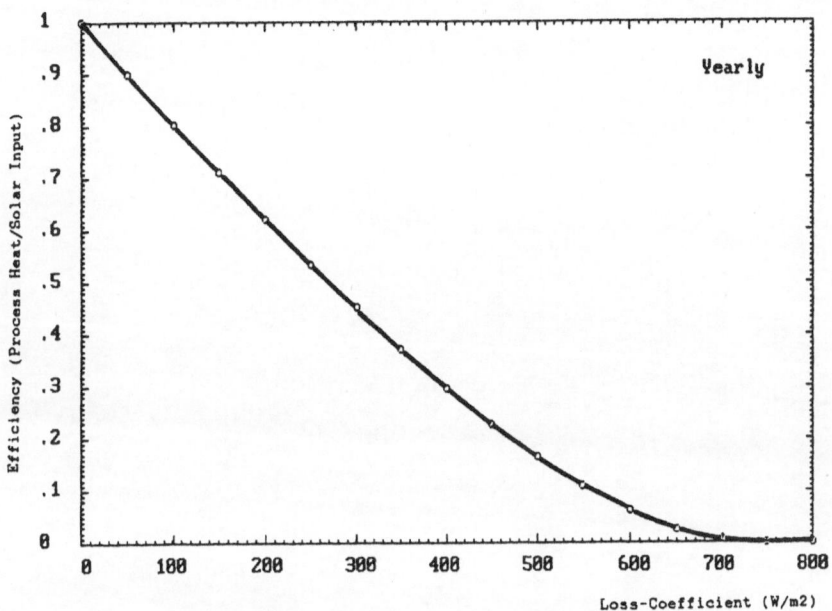

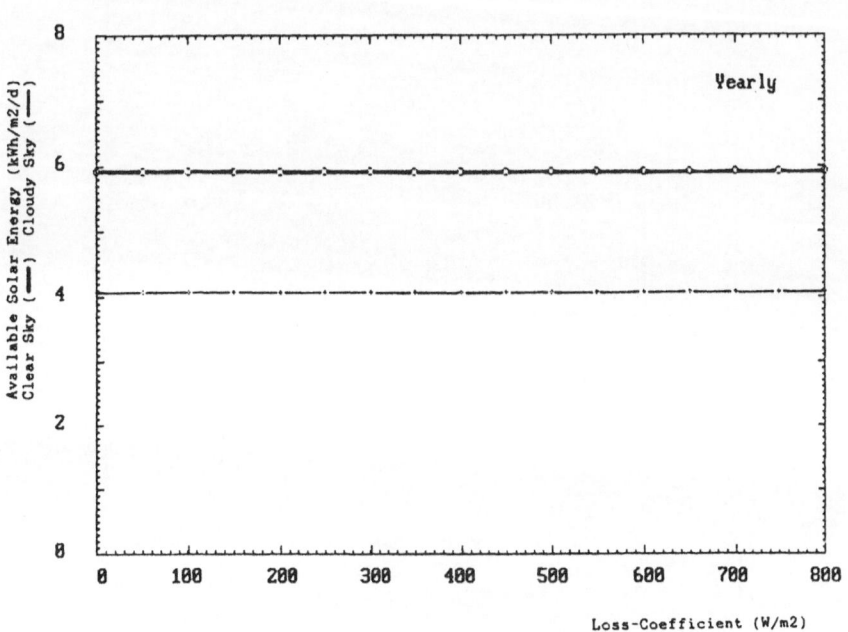

5. Applications

The advantage of the formulation with the reduced loss coefficient V is its general validity for arbitrary cases of the receiver loss function $V_R(T)$ and the optical concentration properties of the heliostat field-receiver aperture design. Its direct significance is the energy flux density of all collected losses. Therefore, V is limited to $E \cdot F_{COS}$ which is the solar flux density accepted by the heliostat field, i.e., the true energy flux input of the solar plant.

5.1 Estimating Receiver Losses

The receiver will loose energy by backscattering, convection, heat conduction into the ambient, and thermal radiation. We estimate the lowest loss possible for process heat consumption in the receiver. Two cases are considered:

- an endothermal reaction driven at constant temperature T (e.g., high throughput of chemicals with little turnover)
- heat extraction where the fluid enters at low temperature T_i, leaving the receiver at high temperature T_o.

In any case the receiver aperture area A_R is chosen to be the smallest possible area, which is the focus plane of beam crossover. We assume all avoidable losses (e.g., inproper absorptance, convection, heat conduction) to be neglegible compared to the unavoidable thermal radiation loss through the very aperture.

In a perfect cavity receiver the radiation energy density is uniform and all the walls are at the same temperature T_R. Then

$$V_R(T_R) = A_R \cdot \sigma \, T_R^4.$$

The fluid temperature T is $< T_R$; the difference $T_R - T$ depends on the efficiency of heat transfer through the reactor walls into the chemical reactive or sensible heat transporting fluid.

In an inperfect cavity part of the thermal emission of hotter wall portions is absorbed by less hot wall areas. In the limit of an open receiver (Fig.2.5) no such reabsorption will occur. We will consider the latter case in more detail to estimate the loss coefficient V of such a receiver.

The heat transfer through the reactor walls is rate determining of the process heat consumption. We assume the extracted process heat (in units W) through wall area dA to follow from

$$dQ_R = k(T_R - T)dA$$

where T_R is the local temperature of the wall exposed to the radiation flux. T is the internal local temperature of the fluid and k the heat transfer coefficient (in units $W/m^2/K$). Along the reactor wall there will be locally different temperatures T_R. We assume a linear temperature increase of T_R from the input side of the fluid (input temperature T_i) up to a higher T_R at the output side (fluid temperature $T_o > T_i$), see Fig.5.1.

The local energy balance equation is

$$\alpha \, E_R \cdot dA = \varepsilon \sigma T_R^4 dA + k(T_R - T)dA + \text{convective losses}$$

α and ε are (temperature dependent) absorptance and emittance of the receiver wall. E_R is the locally homogenous incident radiation flux density on to the absorber. We calculate the local thermal receiver efficiency

$$\eta_R = dQ_R/E_R dA = k(T_R - T)/E_R = \alpha - \varepsilon \sigma T_R^4/E_R.$$

With

$$T_R = T + \eta_R E_R/k$$

we arrive at the implicit relation

$$\eta_R = \alpha - (\varepsilon \sigma/k^4)(kT + \eta_R E_R)^4/E_R.$$

Extending the calculation to a linear temperature rise of the fluid along the flow direction we obtain

$$\eta_R = \alpha - (\varepsilon \sigma/5(T_o-T_i)k^5)((kT_o+\eta_R E_R)^5-(kT_i+\eta_R E_R)^5)/E_R$$

To illustrate the strong dependence of η_R on the heat transfer coefficient k we plot in Fig.5.2 specific examples:

$$\alpha = 1 \qquad \varepsilon = 1 \qquad \text{(perfect black receiver surface)}$$

$E_R = 1000 \, E_{sc}$ (incident radiation concentrated to 1000 times the solar constant $E_{sc} = 1367 \ Wm^{-2}$)

and consider three cases

(i) $T = 1000$ K   (constant fluid temperature)
(ii) $T = 1200$ K   (constant fluid temperature)
(iii) $T_i = 1000$ K   (linear temperature gradient in the fluid
          to   along the absorber length dimension)
          $T_o = 1200$ K

The receiver loss is

$$V_R = A_R E_R - Q_R = A_R E_R (1 - \eta_R)$$

and the reduced loss coefficient is, by noting

$$E_R = E \cdot F_L \cdot F_F \cdot A_F/A_R = E \cdot F_{COS} \cdot C \qquad \text{because of} \qquad C = A_F \cdot F_L \cdot F_F^*/A_R$$

$$V = E \cdot F_{COS}(1 - \eta_R) = \sigma T^{*4}/C.$$

The effective loss temperature T^* depends strongly on the heat transfer through the wall. Values of T^* are listed in Fig.5.2.

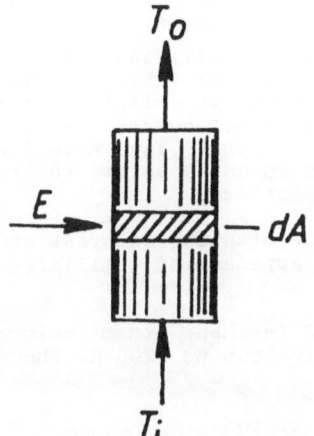

Fig.5.1 Illustration of an open receiver cylinder. The heat transport fluid (whether carrying sensible heat or being endothermally reactive) enters with temperature T_i and leaves with temperature T_o. The marked strip is a differential area dA of local wall temperature T_R and (internally) local fluid temperature T, with $T_i < T < T_o$.

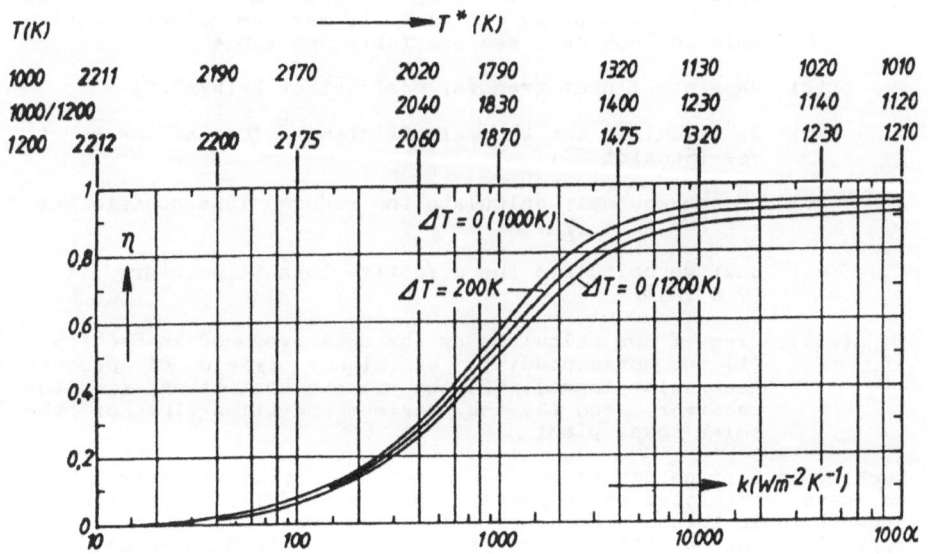

$T(K)$	$\longrightarrow T^*(K)$								
1000	2211	2190	2170	2020	1790	1320	1130	1020	1010
1000/1200				2040	1830	1400	1230	1140	1120
1200	2212	2200	2175	2060	1870	1475	1320	1230	1210

Fig.5.2 Receiver thermal efficiencies versus wall-fluid-heat transfer coefficient k ($Wm^{-2}K^{-1}$). Assumed is a perfect black absorber and a homogeneous incident solar radiation of flux density 1.367 MWm^{-2}. Two curves apply to zero temperature difference between inlet and outlet of the fluid; their temperature levels are 1000 K and 1200 K. The center curve applies to a linear temperature rise in the fluid over the length dimension of the absorber. The corresponding effective radiative loss temperatures T^* (a measure of the external absorber wall temperature) are reported on top of the diagram.

5.2 Applications in Examples

From Fig.5.2 we could calculate the loss coefficients V for various heat transfer values k. We note, however, that we assumed V to be constant in time from start to stop of the plant operation. This is reasonable for any application of process heat with fixed reaction temperature T (or temperature limits T_i, T_o). With increasing solar flux the mass flow rate is adjusted so as to fulfil that condition of constant process temperatures.

Hence, for rendering numerical examples with the material presented in this investigation we take the approach of utilizing the information contained in Fig.5.2.

We consider a reaction at about 860° C (methane steam reforming, LURGI, W.-D. Müller 1986) which is equivalent to 1200 K. The numerical calculations follow the scheme

(i)   a concentration factor C is stated

(ii) we calculate a maximum flux density of concentrated radiation at the receiver $E_R = C \cdot 700$ W/m^2. The value 700 W/m^2 is arbitrary but close to maximum values obtainable in Tabernas, see the Table Section 4

(iii) we state a heat transfer coefficient k (W/m^2/K)

(iv) we calculate the thermal efficiency η_R of the receiver (Section 5.1)

(v)   we subsequently calculate the reduced loss coefficient $V = 700 \ (1 - \eta_R)$

(vi) next we calculate the effective loss temperature $T^* = (V \cdot C / \sigma)^{1/4}$

(vii) from V we calculate by the data produced in Section 4 (1) the corresponding efficiency (yield of process heat/solar input), (2) the process heat taken from the receiver, and (3) the maximum operation time of the solar tower plant.

Such calculations can be performed for any day of the year. Here we have restricted us to yearly averages of daily operation to show the main trends. For easy calculation we utilized quadratic approximations of the various yearly averages in dependence on V (V in units W/m^2)

Yield (Efficiency) $\eta = 1 - 2.060 \cdot 10^{-3}V + 7.840 \cdot 10^{-7}V^2$
Available Process Heat $Q = 4.075 - 8.324 \cdot 10^{-3}V + 3.120 \cdot 10^{-6}V^2$ kWh/m^2/d
Maximum Operation Time $t = 8.169 - 3.816 \cdot 10^{-3}V - 5.376 \cdot 10^{-6}V^2$ hours/d

The results are listed in Table 5.1

Table 5.1 Yields, Available Process Heat, Maximum Operation Time in dependence on optical concentration factor C and heat transfer coefficient k. Reaction temperature 1150 K. Open receiver. Yearly Averages

C	k W/m^2/K	V W/m^2	T* K	Yield	Process Heat kWh/m^2/d	Maximum Operation Time hours/d
2000	10 000	75	1275	0.85	3.47	7.9
	5 000	105	1388	0.79	3.24	7.7
	4 000	122	1439	0.76	3.11	7.6
	3 000	150	1517	0.71	2.90	7.4
	2 000	207	1643	0.61	2.49	7.1
	1 000	343	1865	0.39	1.59	6.2
1000	10 000	121	1208	0.76	3.11	7.6
	5 000	144	1261	0.72	2.94	7.5
	4 000	155	1286	0.70	2.86	7.4
	3 000	175	1325	0.66	2.71	7.3
	2 000	214	1393	0.60	2.44	7.1
	1 000	315	1535	0.43	1.76	6.4
600	10 000	183	1181	0.65	2.66	7.3
	5 000	202	1210	0.62	2.52	7.2
	4 000	212	1223	0.60	2.45	7.1
	3 000	227	1245	0.57	2.35	7.0
	2 000	256	1283	0.52	2.15	6.8
	1 000	333	1370	0.40	1.65	6.3
400	10 000	263	1168	0.51	2.10	6.8
	5 000	278	1184	0.49	2.00	6.7
	4 000	286	1191	0.47	1.95	6.6
	3 000	298	1204	0.46	1.87	6.5
	2 000	320	1226	0.42	1.73	6.4
	1 000	379	1279	0.33	1.37	6.0
300	10 000	343	1161	0.39	1.59	6.2
	5 000	355	1171	0.37	1.51	6.1
	4 000	361	1175	0.36	1.48	6.1
	3 000	370	1183	0.35	1.42	6.0
	2 000	388	1197	0.32	1.31	5.9
	1 000	433	1230	0.26	1.06	5.8
200	10 000	503	1154	0.16	0.68	4.9
	5 000	509	1158	0.15	0.65	4.8
	4 000	512	1159	0.15	0.63	4.8
	3 000	517	1162	0.14	0.61	4.8
	2 000	526	1167	0.13	0.56	4.7
	1 000	550	1180	0.11	0.46	4.3

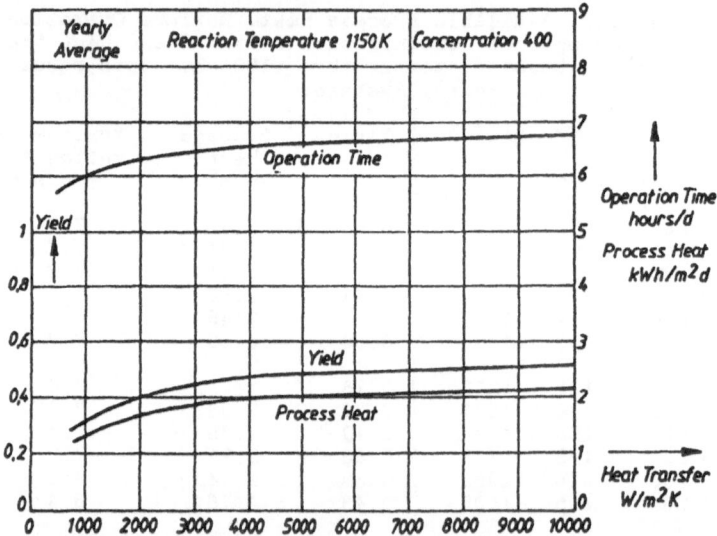

Fig.5.3 Yearly averages of operation time (hours/day), process heat (kWh/m²/day), and yield of process heat from available solar radiative input concentrated 400 times at a fluid temperature of 1150 K in dependence on the heat transfer coefficient k (W/m²/K)

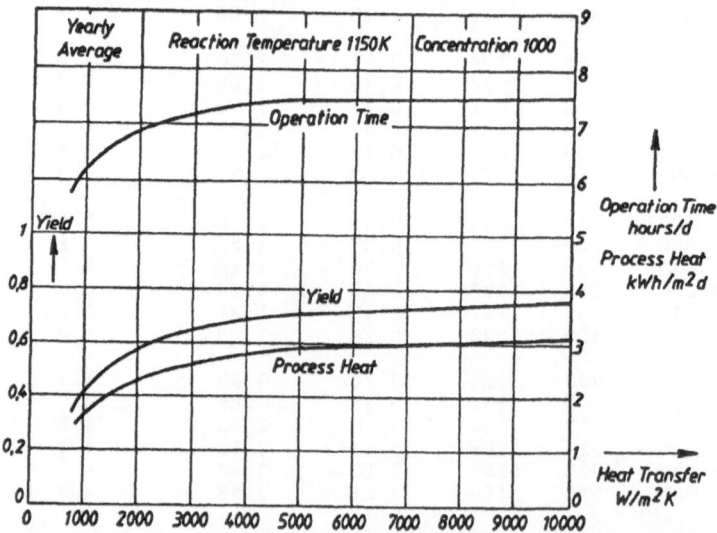

Fig.5.4 Yearly averages of operation time (hours/day), process heat (kWh/m²/day), and yield of process heat from available solar radiative input concentrated 1000 times at a fluid temperature of 1150 K in dependence on the heat transfer coefficient k (W/m²/K)

In Figs. 5.3 and 5.4 the data of Table 5.1 are plotted for the optical concentrations 400 and 1000. Since the receiver aperture becomes smaller with increasing optical concentration, the yield of usable process heat and the operation time rise at a given receiver temperature. Note, however, the assumption made in the present estimate of the monthly and yearly average yields: all the solar radiation is within a cone of half angle ß and can be focused perfectly onto the receiver aperture. An optical concentration

$$C = F_L \cdot F_F^* / (4 \cdot \tan^2 \beta) = 400$$

(see Section 2.5) is achievable with the usual land use factor $F_L = 0.25$, a field factor F_F^* of 0.8 and a beam divergence of ß = 0.6 degrees. To produce an optical concentration of C = 1000 the beam divergence must be reduced to a difficult 0.4 degrees, where the meteorology of the aureole angle becomes an important limiting influence. The alternative to decreasing ß is to improve the optical focusing power of the system by means of a terminal concentrator, Fig.2.6.

A second computation has been performed which directly delivers daily, monthly and yearly yields. A constant fluid temperature T in the receiver has been chosen, together with a heat transfer coefficient k. Then, for a given time dependent irradiance E(t) of the receiver the flow rate of the fluid is always adjusted as to keep T constant. Dependent on the time varying E(t) and a constant k the receiver attains a temperature T_R > T for driving the heat flux Q = k(T_R - T) into the heat transfer fluid. The receiver loss is put Q_{loss} = σT_R^4 or

$$Q(t) = E(t) - \sigma T_R^4 = k(T_R - T).$$

By recursion techniques $T_R(t)$ = $T_R(E(t), k, T)$ is calculated in time steps of minutes. Q(t) is integrated over the day; the daily process heat yields are summed over the months and finally the monthly values are collected to the yearly available process heat.

In Figs. 5.5 and 5.6 results are plotted for two optical concentrations, 400 and 600. In both cases the heat transfer coefficient has been put to 3000 $Wm^{-2}K^{-1}$ and the absorptance (emittance) of the receiver to 100%.

The estimates in Table 5.1 agree well with the present elaborate calculations. The computations have been performed for different fluid temperatures. Demanding an at least 50% yearly average yield of process heat from solar radiative input restricts the fluid temperature to less than 1200 K at an optical concentration C = 400 but allows 1300 K at a concentration of C = 600; for this temperature the yield is already down to 30% at C = 400.

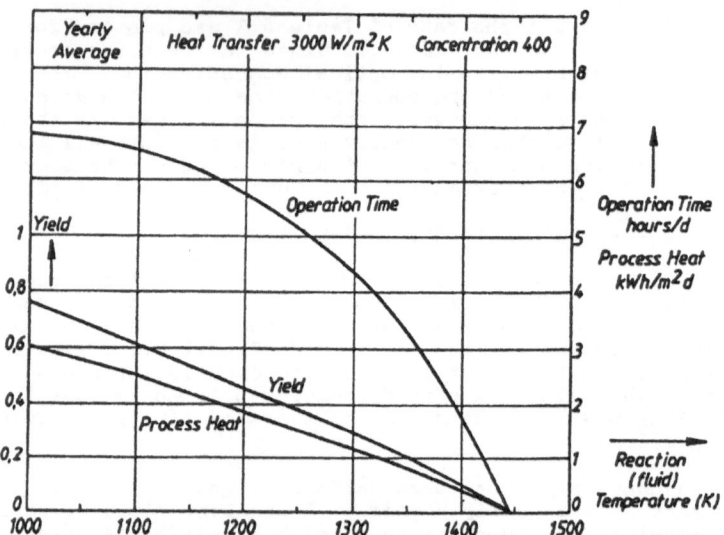

Fig.5.5 Yearly averages of operation time (hours/day), process heat (kWh/m²/day), and yield of process heat from available solar radiative input concentrated 400 times for an heat transfer coefficient k = 3000 W/m²/K in dependence on the fluid temperature (K)

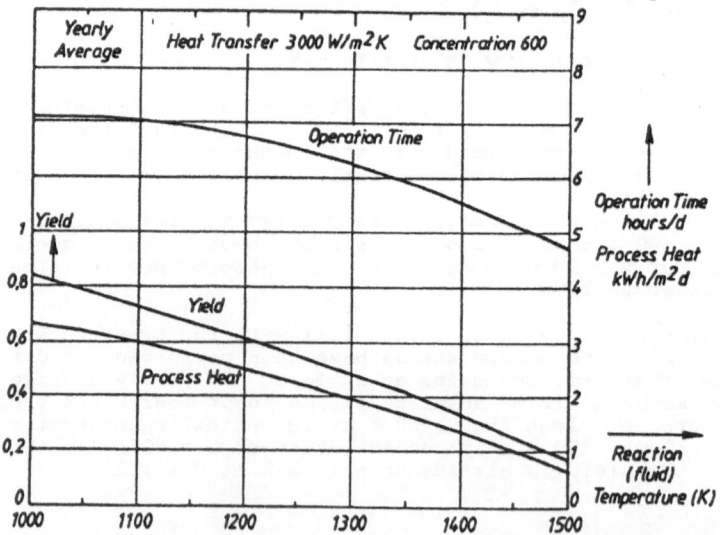

Fig.5.6 Yearly averages of operation time (hours/day), process heat (kWh/m²/day), and yield of process heat from available solar radiative input concentrated 600 times for an heat transfer coefficient k = 3000 W/m²/K in dependence on the fluid temperature (K)

6. Conclusions

The emphasis in the present investigation has been on two topics

(1) The meteorology of the availability of direct solar radiation in daily and seasonal trend has been developed in a modern and versatile manner (Section 3). In particular for the site Tabernas detailed data of the essential meteorological parameters are given. The direct solar radiation has been calculated. The values obtained are in agreement with observations on the spot. Thus, the data can be used for solar radiation flux calculations in any focusing system.

(2) Process heat production and maximum operation times have been calculated with the previously mentioned meteorological input. A formulation is chosen which allows a general description of daily maximum operation time of the plant, yield (efficiency) of process heat related to the available solar input. Monthly and yearly averages of these quantities are presented. The common general variable is a loss coefficient V, which relates the receiver loss V_R to the mainly by field geometry and heliostat properties determined design factor $A_F \cdot F_L \cdot F_F^*$ (A_F, heliostat field land area; F_L, landuse fraction; $F_F^* = F_F/F_{COS}$, with F_F the field factor, F_{COS} the cosine factor). V is taken to be constant in time.

In the present investigation, the field factor has been calculated for a circular field uniformly occupied by heliostats.

Results and conclusions are

- The yield data (yearly averages) for process heat consumption at 1150 K (about 860° C, taken to represent the methane steam reforming reaction) are collected in Table 5.1 in dependence on concentration factor C and coefficient k of heat transfer between the external absorber wall and internal working fluid

- For yields > 60% the concentration factor should exceed 600, the heat transfer coefficient should be greater than 3 000 $Wm^{-2}K^{-1}$

- The concentration factor $C = A_F \cdot F_L \cdot F_F^*/A_R$ (Section 2) contains the receiver aperture area A_R. To capture all of the radiation focused from the heliostats at the rim of the heliostat field into the receiver, the aperture size A_R is by geometry related to the field size A_F. In case of a down facing cavity receiver the concentration factor is essentially $C = F_L \cdot F_F^*/(4 \cdot \tan^2 \beta)$, see Section 2.5, and therefore limited
      (i) by the landuse factor F_L (in combination with F^*) and
      (ii) by the divergence angle β of the solar beam.
Using $F_L = 0.25$, $F_F^* = 0.8$ and $\beta = 0.6$ degrees the concentration factor is of order 500. The SSPS-CRS-installation reports a concentration factor of over 400.

- The k - value of a metal tube - pressurized air receiver structure is at flow rates of 5 m/s near 300 $Wm^{-2}K^{-1}$. A required k > 3 000 $Wm^{-2}K^{-1}$ is possible with steam (5800 $Wm^{-2}K^{-1}$) and sodium (25 700 $Wm^{-2}K^{-1}$), these data taken from Fricker (1983). Utilization of direct absorption is, therefore, of advantage if possible in the engineering of the process in question. Another device worth mentioning for increasing the concentration factor is to improve the focusing power of the system: terminal concentrators (Section 2.5).

- The yield in a 1300 K process rather than in a 1150 K process shows a rapid break down of the efficacy of the installation with optical concentration C = 400. The corresponding numbers are

```
at 1500 K    0.57    2.35    6.3
at 1300 K    0.30    1.25    4.4
```

for yield, available process heat ($kWh/m^2/d$) and maximum operation time (hours/d) in yearly average.

- Operating the receiver with internal fossil fuel firing to keep its output constant over 24 hours would mean that in yearly average at most for 1150 K about 6 hours are assisted by solar input with a contribution of about 2.35 kWh per m^2 of heliostat mirror area at concentration factor of 400 and a heat transfer coefficient of k = 3000 $W/m^2/K$.

7. Recommendations

We divide the recommendations into three paragraphs:

(1) Meteorology
- The analysis presented in this investigation relies partly on data which are not recorded on the spot but at distant sites. In particular orographic cloud formations and their daily and seasonal dependence should be included to improve the description of the local Tabernas meteorology.

- Recordings of cloud structure, stratiform or convective, should be accumulated at Tabernas and Fourier analysed for typical lengths of shadow periods. Such data are important in combination with transients of the installation.

- The seasonal and yearly spread (variance) of the meteorology data should be investigated and included for the forecast of the band width of plant productivity. This remark is also related to the appreciation of solar multiples.

(2) Process heat production
- High process temperatures require a concentration factor beyond the present about 400. Measures should be taken to enhance the concentration by
    (a) increasing the landuse factor (installing more or bigger heliostats on the same land area)
    (b) investigating terminal concentrator structures to increase the optical focusing power.

- High process heat temperatures demand high radiation flux densities at the absorber. The heat transfer coefficient should exceed 3000 $W/m^2/K$: direct absorption, if possible by process engineering, is a recommended solution. Other solutions, e.g., cavity receivers of large internal absorber area are to be considered.

- The reaction temperature should be kept below 1000°C as long as the concentration factor is below 1000, otherwise the solar process heat yield is less than 60% and the maximum available operation time is shorter than 6 hours in the yearly average. This statement implies a heat transfer coefficient of more than 3000 $W/m^2/K$.

(3) General
- In the present investigation several simplifying assumptions were made to keep the final results as free as possible from special structural parameters, e.g., circular heliostat field, maximum power design, rotational symmetry of the receiver. In a next step the particular Tabernas field and its receiver aperture geometry can be fully taken into account. That is rather a matter of computation than of new principles to be included.

- The energy loss function V_R of the receiver (or any forthcoming new receiver design) should be evaluated in dependence on absorber wall temperature. Having $V_R(T)$ available the general loss coefficient V can be determined and by it the process heat and operating time yield calculated. In principle V need not be constant in time, but then its correlation to solar radiation level has to be established. Such detailed information implies individual functions and parameters for any particular receiver.

- A main part of this investigation is related to the solar meteorology at Tabernas. Detailed data in a workable manner are now available. Their application to process heat yield and the analysis of plant operation time has been put forward. The natural final goal of such an investigation is the development of a computer assisted program to evaluate the daily, monthly, yearly solar process heat yield at Tabernas for input data such as working fluid temperature and heat extraction rate. The outcome is paramount to the layout of storage and solar multiple and the basis for designing optimum energy economy/investment.

N.U.TECH GmbH

LITERATURE SURVAY IN THE FIELD OF
PRIMARY AND SECONDARY CONCENTRATING
SOLAR ENERGY SYSTEMS CONCERNING
THE CHOICE AND MANUFACTURING
PROCESS OF SUITABLE MATERIALS

A. GRYCHTA
J. KAUFMANN
P. LIPPERT
G. LENSCH

N.U. TECH, NEUMÜNSTER

Contents   **Page**

1. Introduction

The following report summarizes the actual knowledge concerning aspects of solar energy, that had been specified in the convocation.

To do that, the relevant literature concerning the use of mirror materials and -structures in the solarthermal and combined solarthermal/photovoltaic energy conversion, had been collected, valued and ordered in the following way:

- Elastic mirrors and "Stretched membrane"
- Fresnel- and holographic concentrators
- Structures of concentrators
- Secondary mirrors
- New materials and selective mirrors
- Combined photothermal/photovoltaic energy conversion

Following journals had been taken into consideration:

- Thin Solid Films
- Solar Energy
- Solar Energy Materials
- Applied Optics
- SPIE-Proceedings
- Optical Engineering

So far, the relevant periodicals had been seized.

Since the source materials did not seem to be satisfying in some fields, an investigation at INKA had been done to manifest and complete the collected information.

The periodicals had been looked through for the period from 1980 until August 1986. Because of that, important articles should not have been missed.
Of course some fundamental articles and development reports of the SERI- and SANDIA laboratories had been taken into consideration, too.

This way the presented report pretends to give an overview about the state of the art and the direction of development in some fields of solar energy.
The report is to be a help for future decisions and/or formulations concerning new research and development.

2. Energy conversion and structures of concentrators

To convert solar energy highly efficient, one has to concentrate the solar radiation because of the thermodynamic law. /1-9/
There are two different ways to concentrate radiation:

1. Concentration by reflecting systems
2. Concentration by refracting and diffracting systems

In most applications, solar energy is concentrated by using central or distributed mirror-receiver-systems (CRS, DCS, parabolic dish, parabolic through etc.).

To reach a high efficiency in solar energy conversion, one needs mirror materials which perform a lot of requestion. Some of these are only partially solved and sometimes different requestions even contradict. The following enumeration contains the most important requestions:

- high reflectivity and high specularity vs all of the spectrum of the sun (this is not necessary for some hybrid systems /205 ff/)
- low cost mirror production
- small quantity of material used (thin films)
- easy cleaning of mirrors
- long lifetime
  * high corrosion resistance
  * low photodegradation
- high temperature stability

So one has to find the best adapated metarials and structures for special applications.

In general two different types of mirror structures are distinguishable. These are first- and second surface mirrors. When first surface mirrors are used, the reflecting film is at the side of incoming radiation. While using second surface mirrors, the reflecting film is at the opposite side of the substrate.

The best reflectivity ever observed is that of silver and aluminium films (typical value of hemispherical reflectance is 97% respectively 91%) /8,14,18,19,29/

Since these elements show the highest reflectivity, a lot of alloys, basing on these elements (e.g. Al-Ag, Al-Be, Al-Si, Al-Ag-Cd and even Al-rare-earth-combinations) had been produced and examined. /11,14,18,29/

Aspired results should be the following:
- higher corrosion resistance of silver mirrors
- higher reflectivity of aluminium mirrors

Neither an increase in reflectivity of aluminium mirrrors, nor a longer lifetime of silver mirrors with the same high reflectivity could be achieved. That way the attempt to improve the properties of the mirrors failed. /11,14,18,29/

It seems to be more promising, to protect the mirror from the environment by producing a multilayer with a protecting, highly transparent film at the outside.
These transparent covers are not allowed to react with the reflecting film, as that would lead to a decrease of reflectivity. The adhesion at the reflecting layer has to be high. The abrasion resistance has to guarantee an easy cleaning, too. /20,22,23,26,27,31,32,34-37/

Conventional mirrors are generally designed in the following way: At the backside of a glass substrate (typ. 5mm thick) you find a reflecting film (typ. 100 mm thick) made of aluminium or silver. This film is covered from a protecting film and another stabilizing fibre structure.

To minimize absorption losses float glass with a low iron content is used. /6,8,37/
To realize first surface mirrors that show the highest reflectivity and to protect second surface mirrors well adapted, highly transparent protecting films had been developed. Following materials had already been taken into consideration:

- silicon resins   /31,37/
- teflon   /31,37/
- silicon dioxid   /22,37/
- organosilanes   /22,37/
- sol gel films   /34,35,37/
- PMMA   /2,29,37/
- polymers   /23,37/

Concerning that special part of solar energy, a literature survey had been done in 1985. /37/

One result of this report is, that polymers, PMMA, organosilanes and silicon dioxides are most promising candidates especially when aluminium mirrors are used.

As far as we know, practical experiences in DCS and CRS exist only for second surface mirrors. First surface mirrors had only been evaluated in laboratory testing.

Protecting films to reduce dust accomodation and/or to improve the cleaning procedure had been examined a few times as it is documented in the report mentioned before.

Mirror structures as discribed above are widely used in solar thermal, concentrating conversion systems. Following concentrator structures are distinguishable:

1. Heliostats
2. Parabolic trough
3. Parabolic dish
4. Fresnellenses
5. Holographic concentrators
6. Stretched membrane
7. V-like concentrator
8. Cassegrainian concentrator
9. Test bed concentrator
10. Prism-concentrator
11. Compound parabolic concentrator   (CPC)
12. Compound elliptical concentrator   (CEC)
13. Trumpet like concentrator   (TLC)
14. Compound triangular concentrator   (CTC)
15. Sea shell collector

Generally all kinds of concentrators could be used as primary
concentrators. The concentrators under topic 11 ....15 are
also used as second concentrators in two-stage systems.

This paper mainly deals with the concentrators named under
3,4,5,6,11,12,13 and 14.

2.1. Fresnellenses and holographic concentrators

As an alternative to reflecting systems, refracting and diffracting systems are discussed. Fresnellenses as refracting- and holographic concentrators as diffracting systems are discussed in the literature under the sources /38 ....54/.

The use of holographic concentrators had only been examined in the very last time. The state of the art concerning that field of solar energy is still very low and as far as we know only the Technical University of Aachen and a research laboratory of Polaroid work in this field.

The result of this R & D should be the development of materials that will make solar energy conversion more competitive towards conservative energy conversion systems.

There are one - and two dimensional fresnellenses. (linear and circular fresnellenses). Moreover different kinds of impressed structures are distinguished:

- plane lenses
- roof lenses
- curved lenses

The main part of present development is the improvement of colour-corrected linear fresnellenses.
ENTECH - corporation in the U.S.A. equipped and tested concentrating PV-systems using lenses described above.
Picture 2.1.1 shows that kind of system.

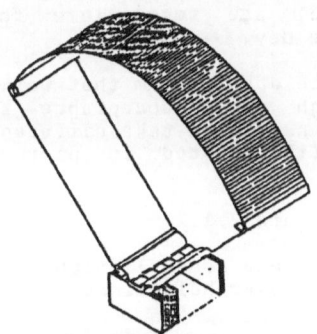

Picture 2.1.1
Fresnellense as a
concentrator in a
PV-system

As in reflecting concentrators it is possible to use fresnellenses in tracking systems, too. Moreover it is possible to install the fresnellenses stationary and track the absorber. /38/
The theoretical limit of linear concentrators is 200. /51/
Until now, concentration ratios up to 90 have been realized.
This ratio is only achievable if the acceptance halfangle is smaller than 5 mrad.

More often concentration ratios between 10 and 40 are reached.
/41,46/

Remarkable is a comparison between parabolic mirrors and
different fresnellenses done by Lorenzo and Luque. They showed
that in a two stage concentrator where the first stage is a
parabolic mirror or a fresnellens, a curved fresnellens leads
to the best results. On the other hand a parabolic mirror
exceeds the easily fabricatable fresnellenses.
Another advantage of fresnellenses is, that they restrict a
tracking precision that may be 2 - 4 times less precise than
comparable mirrors but the part of the useable diffuse light
is extremely reduced.

Also holographic concentrators, performed as lenses or
mirrors, are taken into consideration to convert solar energy.
These structures show very good optical properties but the
main disadvantage of these systems is their solar selectivity.
The most important results had been presented by W. Windeln in
1985. /44/
Diffraction efficiencies of up to 97% are realized for
dielectric, volumetric hologramms. Scattering losses of the
developed hologramms are as small as that of the undeveloped
gelatine films.
The described hologramms have a size of 40cm x 40cm and could
be manufactured in larger geometries, too.
The reproduceability of the films thickness is said to be very
good.

For a wide range of applications, hologramms in which several
lenses are generated or multilayer hologramms are recommended.
A possible hologramm constellation is shown in picture 2.1.2
where the different parts (1,2 and 3) are responsible for
concentration at different times of the day.

Because it is not possible to realize a hologramm that works
in a wide spectral range and with a high angle acceptance at
the same time, several hologramms are needed to take different
sun positions into consideration, (illustrated in picture
2.1.3). /48/

Picture 2.1.2

Possible constellation of
two-layer hologramms

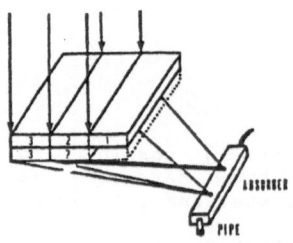

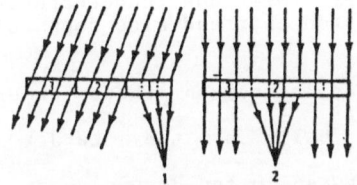

Picture 2.1.3

Juxtaposition of different
hologramms for different
sun position

While Fresnel-concentrators had already been tested outdoors,
holographic concentrators are still in the stage of laboratory
research.

2.2 Stretched membrane

To reduce the production costs of conservative thick glass
mirrors, various instituts started to investigate different
mirror materials and - structures.
Mainly thin glass mirrors and metallized polymers had been
investigated. /60,66,67,75/
The realization of thin and elastic materials offers the
chance to draw up lightweight and cheap concentrators for DCS
and CRS application.
These new structures have to be able to compete with
conventional concentrators not only in the optical and
mechanical properties, but also in the economical aspect.

Stretched membrane concentrators perform these demands if the
mirror materials used are protected against environmental
influences.

The special characteristic of this concentrator is its design,
that also offers large geometries (realized 17m diameter) and
a high quality of the reflecting surface. These concentrators
are designed in the following way:
Two membranes act as the front- and backside of an evacuated
cylinder.
The upper side of the cylinder is covered with a reflecting
material. The variation of the inner pressure leads to a
spherical surface while high concentration ratios are
possible.

The literature /55-76/ describes different types of stretched
membrane concentrators and the materials used.

Moreover there exist some calculations for different kinds of concentrators. /55,56,59,62,70/

At the "Jet Propulsion Laboraties" membrane concentrators of 1m and 2m diameter had been tested that had also been combined to a kind of a parabolic dish.
‾
The combined concentrators had a common focus where a generator had been placed. /69/
These concentrators had been built by' the "La Jet Energy Company".
Concentrators with diameters between 1m and 3m had been built and investigated at SERI.
Hereby silvered polymer films had been used and a specularity of 90% was observeable with a high quality of reflection over a range of wavelength of 400nm to 1000nm and an incident angle of 20° to 60° . /67/

When stretched membrane heliostats of 3m in diameter had been investigated, some deformation problems appeared that are said to be compensateable. /64,68/
The most important advantages of that kind of heliostats are the cheap and lightweight design which leads to a simple tracking system. Cost reductions of up to 50% and weight reductions of up to 75% are foreseen. /73/

Stretched membrane concentrators with 5,2m in diameter had been investigated by Boeing. Aspired limits (max. 1% loss to specularity per year, reflectivity of 93% and life expectance of ten years) could not be reached. Hereby polycarbonate - and polyester films with thicknesses between 0,1mm - 0,2mm had been used.

The University of Texas works in the field of stretched membrane concentrators, too.
By realizing a 9m concentrator, a cheap reflector is to be developed and tested at a test facility of the D.O.E.
As a membrane material fibert-plastic is recommended. The efficiency of that system is discussed under consideration of a peak load during noon. /59/

The largest membrane concentrator, designed as a paraboloid, had been developed in Western Germany. Right now this concentrator is been tested in Saudia Arabia.
Pic. 2.2.1 shows the membrane concentrator designed by "Schlaich und Partner".

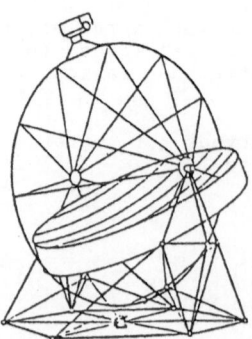

Picture 2.2.1

Stretched membrane concentrator

This concentrator has a diameter of 17m and a membrane made of CrNi-steel with a thickness of 0.5mm. The reflecting surface is realized by using thin glass mirrors of 0.4mm thickness.
Using that concentrator, concentration ratios of up to 2000 (typ. 600) and an electric power of 50 kW are realized.
Long time testing of 1.5 years is to afford further knowledge that should be taken into consideration when future R&D is going to be formulated. /65/

(Private informations included a hint about a membrane concentrator of 38m in diameter and an electrical power of 200 kW that could be developed in Spain).

Summarizing it can be said that it seems to be promising to develope and produce silvered polymeres that show a high reflectivity even for longer times.

These stretched membrane concentrators pretend to be a cost efficient alternative to conventional silver-glass-configurations.
Especially the realization of light and cheap heliostats promises an economical use of these systems. It is worth to mention that activities in this field are not observable in the FRG until now. Moreover large membrane concentrators as paraboloids that use thin glass mirrors should be continiously developed, too.

2.3 Secondary concentrators

Since 1966 a new family of concentrators that had been developed in the USA, the UDSSR and in the FRG at the same time are discussed.
These are the nonimaging concentrators. /2,3,5,82,98,121,124/.
While increasing the divergence, they concentrate radiation, that enters the input aperture and direct it onto the absorber (Pic. 2.3.1)

Picture 2.3.1

Nonimaging concentrator with an acceptance angle of 30°.

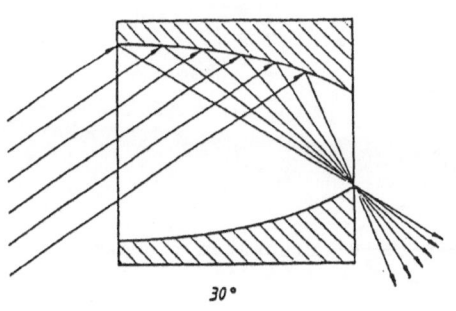

30°

These concentrators are also called ideal concentrators, ideal as all the radiation reaching the input aperture within the acceptance angle also reaches the absorber.
Radiation out of the acceptance angle is reflected.
The most important non-imaging concentrators are:

CPC   (compound parabolic concentrators) /81-85,93-96,98,99,
                                          101,119-121,123,134/

CEC   (compound elliptical concentrators) /82,83,90,98,100,111,
                                           124,133,134/

TL   (trumpet like concentrators)   /87,111,114,127,133,
                                           134/

Picture 2.3.2 shows a CEC, picture 2.3.3 a trumpet like concentrator.

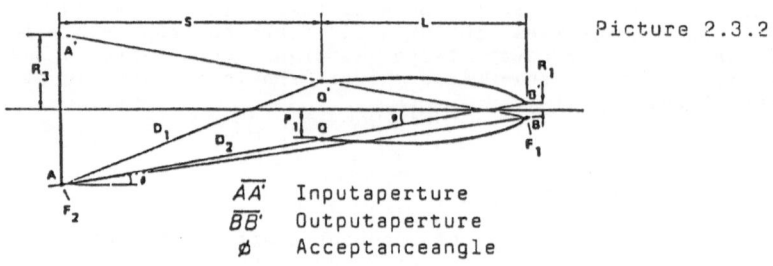

Picture 2.3.2

$\overline{AA'}$ Inputaperture
$\overline{BB'}$ Outputaperture
$\emptyset$ Acceptanceangle

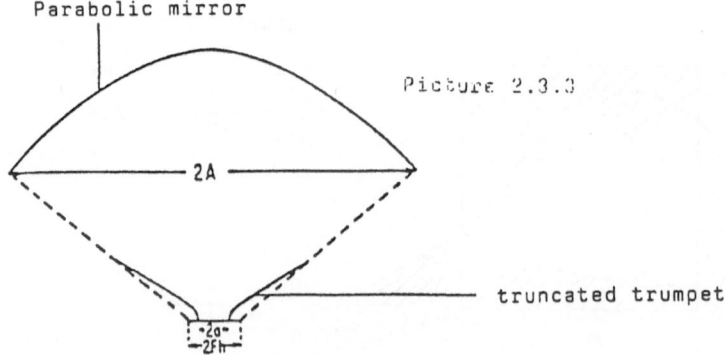

Parabolic mirror

Picture 2.3.3

truncated trumpet

In the very last past a new family of concentrators had been introduced. /112/
These are the CTC´s (compound triangular concentrators). They offer an easier manufacturing process because they don´t have curved, but straight surfaces and they are very small.

As far as we know they are only theoretically discussed in the source mentioned above. As it had been mentioned preliminarly nonimaging concentrators may also be used as primary once.

If high concentration ratios (higher than 10) are required these systems become impracticable because the relation bet-ween input aperture and reflector surface rises incredibly.

An ideal CPC with an input angle of 0.25° and an aperture diameter of 1m had to be 200 m long.

The use of truncated concentrators gives the chance to avoid this disadvantage while a little decrease of concentration has to be accepted. /2,5,83/

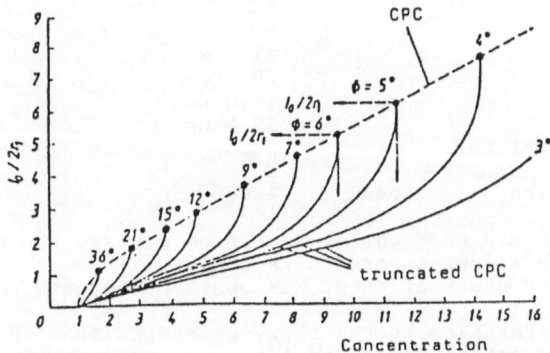

Picture 2.3.4
Decrease of concentration by truncation of a CPC /121,129/.

To realize high concentration ratios in spite of that practical requirements lead to the design of dual-systems. In general these systems are a combination of an imaging first stage (paraboloids, heliostats etc.) and a non-imaging second stage (CPC, CEC,TL etc.).
That way it is possible to design either DCS or CRS applications /90,99,103-105, 111, 114,115,118,134,137/.

Since concentration results from two stages, one can discuss two consequences:

- Realization of concentration ratios higher than 10.000 under use of precise reflector surfaces.

- the required precision of the components may be lowered whereby the costs of heliostats decrease.

Until now only distributed dual systems had been investigated, where shading effects have an important influence.
For each application an optimal truncation may be computed with the required concentration ratio. (See pic. 2.3.4).

The second stage leads to an typical increase of concentration by the factor of 2-4. /93,111/.

Investigations showed that a trumpet structure is to be preferred because of its smallest optical losses. /111,134/
A parabolic-trumpet structure for high temperature use (>400 °C) had been tested at Jet Propulsion Laboratory, with the following data:

paraboloid diameter	11	m
input aperture	76,8	cm
exit aperture	14,0	cm
truncated length	39,5	cm
concentration (trumpet only)	2.1	
over all concentration about	2.000	

The reflecting surfaces are made out of aluminium covered bulk aluminium, respectively silver covered bulk copper.
An increase of generated power of about 20% could be reached.

As the test showed the high energy flux, generated from the first stage requires an active water cooling to prevent the melting or other damages of the secondary mirror.

Moreover the structure has to be tracked precisely to reduce the number of reflections to a minimum (tracking precision smaller than 0.25°). /111/

The results of that investigations showed that it seems to be impossible to use conventional materials because the thermal load required an active cooling that reduces the efficiency of the system.

Annealing effects and temperature variations had not been examined at all.

To solve the material problem it may be possible to use thin films made of thermal and mechanical high resistant materials,that may be deposited on ceramic substrates. Possible candidates are binary and ternary compounds on the base of the d-elements. (compare chapter 3).

Films of the kind that had already been examined generally show a reflectivity like heat mirrors. By variation of process parameters during deposition, it is possible to reach a high reflectivity even in the visible (about 600nm).
These materials have a good chance to be applied in hybrid systems as described in chapter 4.

Very high temperatures (> 1000°C) that lead to an efficient, economical energy production in the MW-range can only be generated in central receiver systems. /2,5,7,134,137/

Because of the small acceptance angle of secondary concentrator structures one can not use a single secondary concentrator at the receiver. To seize all of the heliostate field it is necessary to install several small secondaries in a so called "Fly eye" structure (Pic. 2.3.5). One secondary collects radiation from a special field of heliostats.

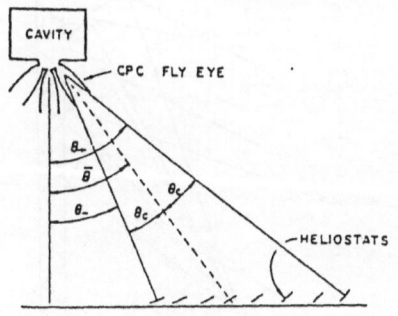

Picture 2.3.5

Central receiver system with "Fly-eye" structure. /2/

An other arrangement that had been suggested by R. Sizmann /7/ is the following:

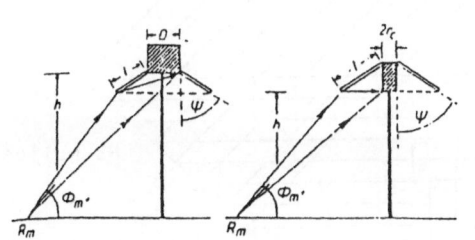

Picture 2.3.6

Absorber placed above, respectively below the secondary mirror

This arrangement is easier to realize and will lead to an increase of concentration by the factor of 2, if the relation l/h is 0.02. /135/
Since a small distance between absorber and generator is desireable, it would be advantageous to place the secondary mirror at a tower and the absorber at the ground.
That kind of solution would also reduce heat losses, because no long pipes are needed.
That way it would also be possible to use highly concentrated solar radiation for chemical processes directly.
Geometrical constellations are shown in picture 2.3.7 /137/ and 2.3.8.

Picture 2.3.7   Picture 2.3.8

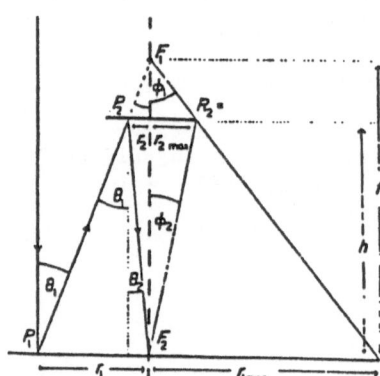   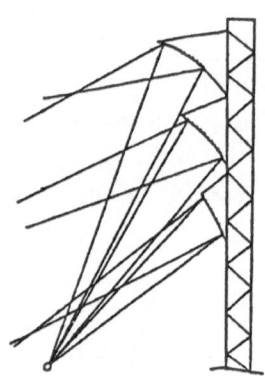

Fundamental experiences have been collected in Odeillo (France) where a parabolic mirror is used as secondary mirror.

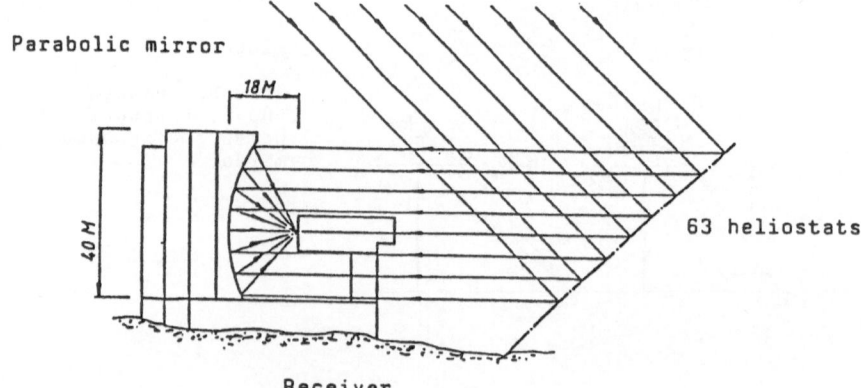

Picture 2.3.9. Dualsystem in Odeillo

At the receiver one can reach temperatures as high as 3800°C at an insolation of 1000 W/m^2 . Hereby it is possible to examine materials as well at very high temperatures as at in wide ranges alternating temperatures.
Summarizing it could be said, that high temperature applications require good care at the choice of material, especially if temperature gradients appear.

It is to be underlined that the fly-eye-structure will cause a high temperature load of the materials used especially if the incoming radiation is not within the acceptance angle of the secondary concentrators.

One can say that all discussed secondary tower reflectors require R & D of useful materials.

3. New materials and solar selective mirrors

A transparent heat mirror (solar selective mirror) is defined as an arrangement with a high transmission (>70%) in the visible and high reflection in the infrared (> 1000 nm) range. In general four types of heat-reflecting films are distinguishable /156,157/:

- metal films, thin enough to be transparent
- metal multilayers
- highly doped semiconductors
- thin films of conducting microgrids

The following types had been investigated:

- thin films of Au and Ag
- Ti O$_2$ / Si O , ZnS/MgF-layers /170/
- Ti O$_2$ / ITO-layers /173,191,203/
- doped Sn$_2$ O$_3$-layers /190,197/
- moreover compounds as CeO$_2$, La$_2$ O$_3$, ZnS, CeF.

All layers like that show both advantages and disadvantages (high costs, unacceptable optical properties etc.)
This chapter deals with heat-reflecting films based on compounds of d-elements.

These compounds show extraordinary properties:

- high mechanical resistivity
- high corrosion resistance
- high melting point.

Compounds like TiN, HfN and ZrN are used as protecting coatings for tools.
These coatings have been deposited using PVD-and CVD-processes to increase the tribological properties and lifetime of the materials drastically.   /147,155,161,162,163,165,180,183,-187,192/

We collected the important data /155/ of some ceramics in the following table:

Table 3.1

phase	density (g/cm³)	melting point °C	hardness (HV)	E-modulus kN/mm²	spec. el. resistivity μΩcm	thermal exp. coeff. 10⁶ K⁻¹
TiB₂	4 50	3225	3000	560	7	7.8
TiC	4 93	3067	2800	470	52	8 0 - 8.6
TiN	5.40	2950	2100	590	25	9.4
ZrB₂	6.11	3245	2300	540	6	5.9
ZrC	6 63	3445	2560	400	42	7.0 - 7.4
ZrN	7.32	2982	1600	510	21	7.2
VB₂	5.05	2747	2150	510	13	7.6
VC	5.41	2648	2900	430	59	7.3
VN	6.11	2177	1560	460	85	9.2
NbB₂	6.98	3036	2600	630	12	8.0
NbC	7.78	3613	1800	580	19	7.2
NbN	8.43	2204 z.	1400	480	58	10.1
TaB₂	12.58	3037	2100	680	14	8.2
TaC	14.48	3985	1550	560	15	7.1
CrB₂	5.58	2188	2250	540	18	10 5
Cr₃C₂	6.68	1810	2150	400	75	11.7
CrN	6.12	1050	1100	400	640	(2.3)
Mo₂B₅	7 45	2140	2350	670	18	8.6
Mo₂C	9 18	2517	1660	540	57	7 8 - 9.3
W₂B₅	13 03	2365	2700	770	19	7.8
WC	15.72	2776	2350	720	17	3 8 - 3.9
LaB₆	4.73	2770	2530	(400)	15	6.4

TiN-films show the best tribological properties, they are the best investigated ones and the mostly used ceramics at all.

When sputter-deposition is used, variation of process-parameters like:

- partial pressure of the reactive gas
- substrate temperature
- deposition rate

leads to films with different properties like:

- adhesion
- hardness
- electrical resistivity
- optical reflectivity and transmission.

/140,145,149,150,153,159,161,164,165,181,200/

The gold character, as high reflection in the IR-range, makes films like that interesting to be investigated concerning application's in solar technology.

We point out, that compared with the mechanical properties investigations of the optical properties are still very rare. Only in the last time activities in that field could be registered.

Tables 3.2 and 3.3 list the results of some optical investigations.

Filmmaterial	Source	Spectra	Commentary
TiN_xO_y	189 compare 196		Near normal spectral reflectance for bias sputtered $TiN_x(O_y)$ layers on polished copper. a: $P(N_2)=0.04$ Pa, $P(air)=0$, $V(bias)=-100$ V b: $P(N_2)=0.04$ Pa,$P(air)=1.3 \cdot 10^{-3}$Pa, $V(bias)=35V$ c: $P(N_2)=0.04$ Pa, $P(air)=0$ $V(bias)=0$
ScN_x	184		Reflectivity spectra of ScN_x 1: $P(N_2)=0.045 \cdot 10^{-4}$Torr Filmthickness 420nm 2: $P(N_2)=0.050 \cdot 10^{-4}$Torr Filmthickness 660 nm 3: $P(N_2)= 0.o55 \cdot 10^{-4}$Torr Filmthickness 600 nm
TiN_x TiN_xC_y ZrN_x ScN_X	184		Reflectivity spectra of different materials 1: TiN_x 2: TiN_xC_y 3: ZrN_x 4: ScN_x

Table 3.2 : Optical properties of some hardcoatings

Filmmaterial	Source	Spektra	Commentary
TiN	188		The near-normal specular reflectance and trans-mittance for magnetron sputtered semi-transparent TiN-films on glas. Thicknesses: a = 41 nm, b = 22 nm, c = 14 nm, d = 11 nm
TiN	188		The near-normal specular reflectance and trans-mittance for rf sputtered semi-transparent TiN-films on fused silica. Thicknesses: e = 37 nm, f = 26 nm, g = 13 nm
TiN ZrN HfN	171 compare 154,168,169		Spectral reflectance of CVD, ZrN and HfN and of sputtered TiN.

Table 3.3 : Optical properties of some hardcoatings

Summarizing we can say that the materials presented show a sufficient reflectivity to be used as heat mirrors. ScN shows the best optical data. TiN may be prefered for wide spread application as it is more available than other ceramics.

The high melting point, great hardness, chemical resistance, and high reflectivity promise the described materials to be used in high temperature applications.

Specially tailored hybrid-structures for total energy conversion (photovoltaic and photothermal) may be an application for hardcoatings because of their adaptivity to solar cells. Simultaneous use of conducting hardcoatings that cover a photo cell /141,144/ and work as a heatmirror is possible.

If it is possible to move the edge of reflectivity of TiN into the direction of smaller wavelengths by variation of process parameters, it may be a possible substitute for gold in heatmirrors.

Especially the multitude of variable process parameters causes that a sufficient knowledge about the optical properties of the described hardcoatings does not exist.

4. Combined photovoltaic / photothermal energy conversion

In the field of solar energy a hybrid system is an arrangement that realizes solarthermal and photovoltaic energy conversion at the same time.

There are two ways to reach cogeneration of electrical and thermal energy.
First of all it is possible to use fresnellenses or parabolic (reflecting) concentrators to focus radiation on specially adapted solar cells, to reduce the number of (expensive) solar cells, needed.
As it is not possible to convert all of the radiation into electrical power (cutting wavelength = 1100 nm /212/) the solar cell will be heated.

To maintain the solar cells efficiency, it is necessary to make a cooling available. /213,216,222/

Picture 4.1 and 4.2 show principle arrangements of a hybrid system, whereby the first concentrator is a fresnellense respectively a parabolic trough.

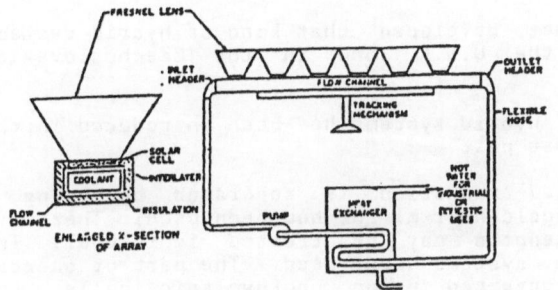

Picture 4.1:

Hybridsystem with
fresnellenses

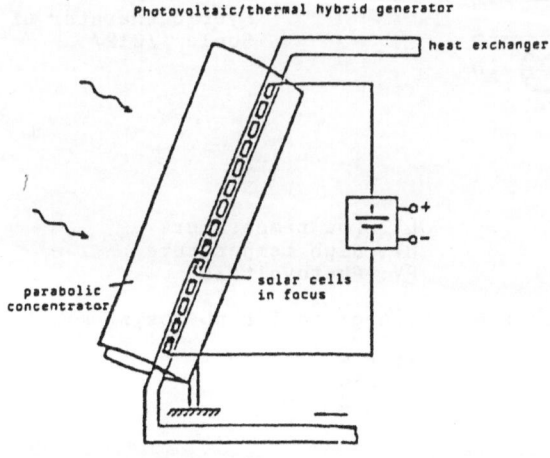

Photovoltaic/thermal hybrid generator

Picture 4.2:

Hybridsystem with
parabolic trough
(HELIOMAN)

Hybrid systems that use a parabolic mirror for concentration had been designed and tested in the FRG by M.A.N. and AEG.

The result was, that systems like that may be realized, even in a higher number of pieces, but they are not economical at all. /213/

Other institutes that developed that kind of hybrid systems are to be found in the U.S.A. and in the Czechoslovakia. /50,211/

A new concept for hybrid systems had been introduced in the U.S.A. in the very last past.

Hereby the concentrated radiation is seperated by using a selective mirror (a gold-heat mirror had been used). That way, two ranges of wavelengths may be treated individual. Two different conversion systems are used. The part of shorter wavelengths will be converted using photovoltaic cells, the part of longer wavelengths will be reflected to an absorber and generate thermal energy that way.

The solar cells may also be cooled, whereby another part of low temperature is utilized. /212,229/

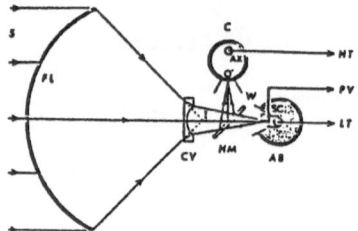

Picture 4.3:

Hybridgenerator of Soule /212/

HM: Heatmirror
FL: Fresnellense

LT: Low temperature
HT: High temperature
PV: Photovoltaic

The following efficiencies had been given for that system:

- HT (150°... 250° C)   17.8 %
- PV (12 V)   9.5 %
- LT (50°... 70° C)   41.9 %

The overall efficiency is 69%.

This is the highest efficiency ever reached in solar energy conversion systems.

It has to be noted, that most of the generated energy is of low temperature, that requires optimal adapted chemical processes to use that energy with acceptable efficiencies.

As an alternative to the presented decentral hybrid systems, it is possible to design a central hybrid system.
An imaginable arrangement could be the following:
Solar cell panels, that are covered with a hardcoating like TiN, HfN or other ceramics, act as heliostats. The hardcoating takes over several functions like:

- heat mirror
- protecting coating for the solar cell
- conducting film for the solar cell

At the tower, an adapted receiver has to be installed, to convert the reflected part of radiation. It would also be possible to realize a secondary concentration, either with the absorber at the tower or with the absorber at the ground.

Hardcoatings promise to solve the problems, that appear when secondary concentration is aspired.

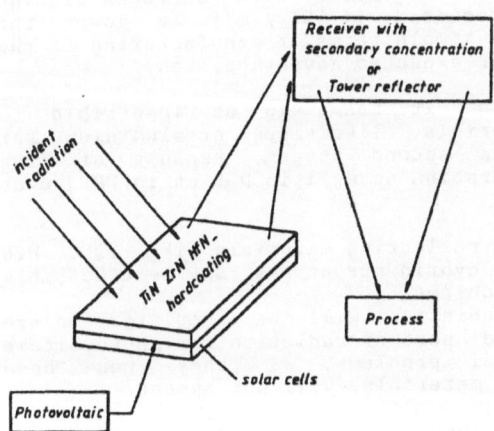

Picture 4.4: Sketch of a central hybrid system

5. Summary

This literature survey reports about:

- fresnellenses
- holographic concentrators
- stretched mebrane
- secondary concentrators
- new materials and selective mirrors
- photothermal / photovoltaic energy conversion

Summarizing we can say, that the further use of fresnellenses for applications in solar energy conversion is not promising.

On the contrary, holographic concentrators may substitute circular fresnellenses, particular, if holographic materials will be resistant against the influencesof the enviroment. At present, these materials are not ready for operation.

The use of stretched membrane concentrators in large scale is highly promising if it is possible to produce membrane helio-stats in a great number of pieces with surfaces of high quality and as cheap as predicted. (< 50$ / m^2). To lower the costs by light design and simple terms of manufacturing of the reflecting films is another expected advantage, too.

Experimental data verify that it seems to be impossible to employ conventional materials, like silver or aluminium, for high temperature uses as second stages, because of the problems caused by absorption as well in DCS as in CRS power systems.

The use of hardcoatings as reflecting materials like TiN, HfN and ZrN, especially in hybrid structures, may be a possible solution to solve those problems.
The reflected part of radiation will be used for solar-thermal-, the transmitted part of radiation for photovoltaic energy conversion. Thermal problems, as they had been mentioned for conventional materials, will not appear.

A high melting point and spectral selectivity are properties, that may lead to a realization of dual systems with secondary concentration.

Particularly we want to point out, that variation of process parameters during deposition may lead to films, optimal tailo-red in reflectivity for application in combined photovoltaic / photothermal energy conversion systems.

6. Literature

230 Sources had been taken into consideration for this literature survey.
Articles that pretended to be interesting had been cited concerning the abstracts.

Every chapter had been collected alphabetically. Following topics had been chosen:

- Summerizing articles
- Conventional mirror materials
- Fresnellenses and holographic concentrators
- Stretched membrane
- Energy conversion and structures of concentrators
- New materials and solar selective mirrors
- Combined photovoltaic/photothermal energy conversion

Summerizing articles

1. Annual evaluation report SERI, several authors,
    SERI/PR - 253 - 2188, (1984)

2. Concentrating collectors, (from: Solar energy technology handbook)
    Rabl, A., Dickinson, W.C., Cheremisinoff, P.N., Solar energy technology handbook, pp 257-343, (1980)

3. Concentrators: a review,
    Dang, A., Energy conversion and management Vol. 26, (1986)

    Abstract:

    Solar energy concentrators including all existing concentrator types are reviewed. The specific aspects covered are design, fabrication, optical and thermal performance, testing, tracking, non-tracking and material requirements. A comparative study of the presently existing forms of concentrators has also been attempted. Some improvements to enhance the performance of the systems have been suggested.

4. Industrial solar thermal concentrators,
    Touchais, M., L'Energie Solaire et sa Maitrise Industrielle
    (1979)

    Abstract:

    The conversion and concentration of solar thermal

radiation are treated, and various industrial solar
heaters and concentrators are presented. The principles
of solar absorbers are discussed, and the concentration
of solar radiation by means of lenses, mirrors and mixed
systems is examined. High-temperature industrial solar
thermal heaters and concentrators developed prior to the
present energy crisis are discribed, including systems of
conical and cylindrical mirrors, paraboloid mirrors,
cylindrical-parabolic mirrors and mirror fields. Current
realizations, plans and possibilities for solar thermal
electricity conversion are presented, including heliostat, concentrator and high-temperature insolator
systems, and the experimental investigation of solar
heater technilogy is discussed. Attention is also given
to the optical properties of parabolic mirrors, the
fabricatons of Fresnel lenses, and the utilization of
optical waveguides and light ducts in solar systems.

5. Nutzung der Sonnenenergie: Konzentrierende Kollektoren
    Fricke, J., Borst, W., Physik in unserer Zeit Nr. 2
    (1981)

6. Overview of solar reflector applications, materials, and
    research and development,
    Call, P.J., Thin Solid Films 72, (1980)

    Abstract:

    Reflector materials with high solar reflectance, long
    lifetime, low cost and appropiate specularity are
    desirable for all concertrating solar collectors ranging
    from augmented flat plates to point focus dishes. A
    review of the cost, lifetime and specularity requirements
    for a variety of systems will be presented. The
    properties of available commercial and advanced
    developmental materials will be juxtaposed with these
    requirements to derive critical reflector research and
    developpment needs. The prospects for inexpensive
    Ag/polymer and thin aluminum reflectors will be
    discussed. Experience to date with commercial
    wet-chemical- processed Ag/glass mirrors in solar
    applications does not inspire a high degree of confidence
    in this baseline mirror system for long lifetime. In
    particular, field exposure at solar thermal facilities
    around the world and, most recently, tests associated
    with qualifying mirrors for the 10 MW Barstow solar
    thermal pilot plant indicate that great care must be
    taken to minimize corrosion of the silver layer. The
    symptoms and causes of this problem as they are presently
    understood will be described and research under way to
    provide more stable Ag/glass mirror systems will be
    reviewed.

7. Solar Energy '85, Burke, W.R., ESA Publications Division,
    ESTEC, Noordwijk, The Netherlands, ISSN 0379-6566, (1985)

8. Solar mirror materials: Their properties and uses in solar concentrating collectors,
    Pettit, R.B., Roth, E.P., SAND 79-2190, (1980)

9. Thin films in photothermal solar energy conversion, Seraphin, B.O., Thin Solid Films 90, (1982)

    Abstract:

    Thin films play an important part in photothermal solar energy conversion. They intercept, redirect and concentrate the incident solar radiation and convert it into useful heat. The cost of materials and their processing makes thin films the most cost-effective solution, since most optical interactions occur within 1 μm of the front surface. However, the large surface-to-volume ratio of thin films faciliates degradation processes that limit the service-life. Durability at elevated temperatures is essential and sets solar thin film technology aside from conventional applications. Spectrally selective coatings are discussed in detail, with emphasis on single-film converters, absorber-reflector tandems, dark mirrors and heat mirrors. A discussion of reflecting films and glazings concludes the review.

Conventional mirror materials

10. Abrasion resistant polymer reflectors for solar applications Assink, R.A., Solar Energy Materials 3, (1980)

11. Al-Ag alloy films for solar reflectors,
    Adams, R.O., Nordin, C.W., Thin Solid Films 72, (1980)

    Abstract:

    Films of Al-Ag alloys were formed using triode sputtering. Films with a wide variety of composition were produced and evaluated. Films deposited at low substrate temperatures have a high specular reflectance. At higher temperatures, two-phases alloys form which have rough low reflecting surfaces.

12. Characterization of heliostat corrosion,
    Shelby, J.E., Vitko, J., Farrow, R.L., Solar Energy Materials 3, (1980)

13. Chemical vapour deposition and thermal energy conversion,
    Haygarth, J.C., Thin Solid Films 72, (1980)

14. Co-sputtering of aluminium-silver alloy mirrors for use as solar-reflectors,
    Adams, R.O., Nordin, C.W., Fraikor, F.J., Thin Solid Films 63, (1979)

15. Deterioration of the silver/glass interface in second surface solar mirrors, Burolla, V.P., Solar Energy Materials 3, (1980)

16. Durability of silver-glass mirrors in moist acid vapors, Coyle, R.T., Barrett, J.M., Call, P.J. Solar Energy Materials 6, (1982)

17. Effect of annealing on the reflectivity of silver films, Vijayakumar, K.P., Purushotaman, C., Thin Solid Films 82, (1981)

18. Optical properties of disordered rare earth-aluminum alloys Trotter, D.M., Solar Energy Materials 3, (1980)

Abstract:

An aluminum surface has a solar reflectance 0.91. The Drude model predicts a reflectivity 99% for Al because it neglects interband transitions which arise because of long range translational order in the pure material. By alloying Al with small amounts of gadolinium and samarium, which are large atoms with the same valence (+3) as Al, materials which are still largely Al but lack long range order are created. The reduction in interband transition strength concomitant with reduction in long range order raises the reflectivity of the alloys relative to pure Al, but only in a fairly narrow wavelength range. Over the rest of the solar spectral region the increased Druderelaxation rate ($\hbar/\tau_o$) resulting from the disorder depresses the refelctivity of the alloys so that a smaller net solar reflectance results. Specular alloy samples are prepared by simultaneoesly evaporating Al and a solute onto chilled (T 100K) substrates. Electron microscopy shows that for solute concentrations <10% the alloys consist of small crystallites embedded in a matrix of disordered material, with the crystallites growing smaller and less organized as the solute concentration increases. Alloys with =10% solute are totally disordered and are characterized by $\hbar/\tau_o$ =0.7 eV, similar to that of liquid Al, and an interband transition strength about 1/3 that of pure Al. Solute concentrations >20% are required to reduce the interband strength to zero; this increases $\hbar/\tau_o$ to = 2 eV. For pure Al $\hbar/\tau_o$ = 0.14 eV.

19. Optical thin films produced by nonvacuum techniques, Haisma, J., Applied Optics Vol.24 No. 16, (1985)

20. Optimization of transparent heat mirrors based on a thin silver film between antireflection films, Köstlin, H., Frank, G., Thin Solid Films 89, (1982)

21. Physical and chemical aspects in the application of thin
    films on optical elements,
    Guenther, K.H., Balzers-Bericht, Liechtenstein, sonst
    keine Quellenangabe möglich!, (ca. 1983)

Abstract:

First, an introductory review of deposition dependent
microstructural peculiarities of real optical thin films
is presented. Then, some physical and chemical properties
of particular interest for the user of coated optical elements are surveyed. Of the optical properties, the refractive index is discussed in detail, and reflectivity,
light scattering and surface plasmon excitation are also
included. Of the mechanical properties, adhesion, hardness, abrasion as well as some aspects of intrinsic
stress are illustrated with practical examples. The importance of the thermal stability of optical coatings is emphasized, and some examples of severe alterations of coating properties after thermal treatment are given. The
discussion of diffusion processes within some coatings
and in the surface region of substrates leads to a survey
of selected chemical reactions between the substrates and
the coating, within the coating itself and between the
coating and the environment. Finally, the degradation of
coatings as a result of radiation impact is discussed
shortly.

22. Plasma-polymerized organosilanes as protective coatings
    for solar front-surface mirrors,
    Bieg, K.W., Wischmann, K.B., Solar Energy Materials 3,
    (1980)

23. Polymer glazings for silver mirrors,
    Neidlinger, H.H., Schissel, P., SPIE Vol. 562, (1985)

24. Progressive changes in microstructure and composition du-
    ring degradation of solar mirrors,
    Daniel, J.L., Coleman, J.E., Solar Energy Materials 3
    (1980)

25. Properties of optical film materials,
    Ritter, E., Applied Optics Vol. 20 No. 1, (1981)

26. Protective coatings for silvered reflectors for solar
    application,
    Wong, S.M., Goggin, R., Call, P.J., Thin Solid Films 83,
    (1981)

27. Protective dielectric coatings produced by ion-assisted
    deposition,
    Sainty, W.G., Netterfield, R.P., Martin, P.J., Applied
    Optics Vol. 23 No. 7, (1984)

28. Reflectance and aging studies of heliostat mirrors,
    Taketani, H., Solar Energy Materials 3, (1980)

29. Reflectivity of silver- and aluminum-based alloys for so-
    lar reflectors,
    Hummel, R.E., <u>Solar Energy Vol. 27 No. 6,</u> (1981)

    Abstract:

    Numerous attempts have been made over this years to
    improve the corrosion resistance of silver and the reflec-
    tivity of aluminium by using alloying techniques. This
    paper discribes the current understanding of the limita-
    tions of these attempts, based on the electron theory of
    metals. Experimental optical data on alloys that utilize
    silver and aluminium as based metals are summarized. It
    is shown that: (1) alloying of silver and aluminium
    reduces the solar-weighted reflectance, (2) solute addi-
    tions may slightly improve the corrosion resistance of
    silver but in turn reduce the reflectance, (3) the para-
    meters which affect the reflectance of a pure metal in-
    clude kind and number of solute atoms, surface reactions,
    surface conditions, kind of mirror preparation and inter-
    band transitions.

30. Scattering defects in silver mirror coatings,
    Pellicori, S.F., <u>Applied Optics Vol. 19 No. 18,</u> (1980)

31. Silicone resins for protection of first surface reflec-
    tors,
    Dennis, W.E., McGee, J.B., <u>Solar Energy Materials 3,</u>
    (1980)

32. Thin film deposition using Sol-Gel technology,
    Puyane, R., Gonzales-Oliver, C.J.R., <u>Battelle, Geneva
    Research Centres,</u> (1981)

33. Thin film solar reflectors,
    Griffin, R.N., <u>Solar Energy Materials 3,</u> (1980)

34. Thin oxid films on glass substrates by a Sol-Gel
    technique
    Puyane, R., Kato, I., <u>Battelle, Geneva Research Centres,</u>
    (1982)

35. Use of Sol-Gel thin films in solar energy applications,
    Pettit, R.B., Brinker, C.J., <u>SPIE Vol. 562,</u> (1985)

36. Zinc oxide multilayers for solar collector coatings,
    Brett, M.J., Parsons, R.R., Baltes, H.P., <u>Applied Optics
    Vol. 25 No. 16,</u> (1986)

37. Literaturstudie zum Thema: Staub auf Spiegeln,Teil A u. E
    <u>FH-Wedel, Bericht 7/85,</u> Susemihl, J., Lensch, G., Brudi,
    K., Lippert, P., (1985)

38. A fixed Fresnel-lens with tracking collector,
Kritchman, E.M., Friesem, A.A., Yekutieli, G.,
Solar Energy Vol. 27, (1981)

39. Color-corrected Fresnel lens for solar concentration,
Kritchman, E.M., Optics Letters Vol. 5 No. 1, (1980)

40. Comparison of Fresnel lens and parabolic mirrors as solar
energy concentrators,
Lorenzo,E., Luque,A., Applied Optics Vol.21 No.10 (1982)

Abstract:

This paper compares the gain that can be achieved with a
one-ot two-stage concentrator, when the first stage is a
Fresnel lens or a parabolic mirror, as a function of the
luminosity of the concentrator. The results show that the
achievable gain using a parabolic mirror is a greater
than that obtained using a flat or roof lens but is lower
than that obtained using a curved lens.

41. Convex Fresnel-lens with larges grooves,
Kritchman, E.M., Friesem, A.A., Yekutieli, G., Solar
Energy Vol. 27, (1981)

Abstract:

The concentration cabability of linear Fresnel lenses is
limited mainly by coma and chromatic aberations and by
the width of the individual grooves of the lenses. We
consider new designs for these lenses in which the concentration is almost independent of the groove s size, and
which are free of coma at the edge of the acceptance
field. Specifically, a concentration factor as high as 34
is achieved for monochromatic radiation coming from an
acceptance angle of + 1 .

42. Design of one-axis tracked linear Fresnel-lenses,
Lorenzo, E., Minano, J.C., Solar Energy Vol. 36, (1986)

Abstract:

The most frequently used criterion in the design of linear Fresnel lenses consists of minimizing the radiation
spread at the collector when the lens is illuminated by
rays that are contained in the full lens acceptance angle
and are incident on the lens in the plane perpendicular
to the lens axis. This paper analyzes this topic and con-
cludes that when the lenses are one-axis-tracked this
criterion can be improved by replacing the perpendicular
plane with another forming a certain angle to it. When
calculated for a specific lens, at the location of Ma-
drid, the new criterion was found to yield an increase of
more than 7% in the total annual energy collected.

43. Development of a stand-alone linear Fresnel lens photovoltaic collector array,
McDanal, A.J., <u>Sandia Report Sand 84-055G p. 117 ff,</u>
(1984)

44. Development of high efficiency holographic solar concentrator,
Windeln, W., Stojanoff, C.G., <u>SPIE Vol. 562,</u> (1985)

Abstract:

In this report, we present the experience gained up-to-
date in the development of holographic solar concentrators. The techniques used in the generation of high efficiency dielectric volume-holograms of the transmission
type are presented in detail. These techniques facilitate
the manufacturing of holographic lenses with diffraction
efficiency in the order of 97%. In order to achieve the
high efficiency, the research team has developed sensi-
tizing and film development procedures for dichromated
layers whose scattering losses are comparable to those of
the unexposed gelatin layer. The manufacturing of the dichromated gelatin layers is performed in-house (30 x 40
cm) and can easily be extended to large apertures. The
layering procedure is a continuous process and is limited
at present only by the travel of the motor-driven table
top. The reproducibility of the film-thickness for a
batch of manufactured 30 x 40 cm^2 holographic plates is
better than + 1 m. The film-thickness variation of the
gelatin film averaged over the entire surface of a holographic plate is in the order of 0.2 $\mu m/cm$.

Theoretical and experimental results are presented for
some relevant parameters that control the diffraction
efficiency of the concentrator. Emphasis is placed on the
problems encountered when a multiple lens-system (stack)
is generated in a single gelatin layer or in an integrated multi-layer hologram.

45. Fresnel lens analysis for solar energy applications,
Lorenzo, E., Luque, A., <u>Applied Optics Vol. 20 No. 17,</u>
(1981)

46. High temperature solar collector with optimal concentration: non fucussing Fresnel lens with secondary concentrator,
Collares-Pereira, M., <u>Solar Energy Vol. 23 No. 5,</u> (1979)

Abstract:

A solar thermal collect is discribed which consists of a
line focus Fresnel lens plus second stage concentrator of
the CPC type. The tracking axis runs north-south, with
fixed tilt equal to latitude. The geometric concentration

is 16, and the acceptance angle 2 theta is 60° ; this allows for large contour and tracking errors and permits collection of most or all of the circumsolar radiation. The receiver, coated with black chrome, is nonevacuated. The collector is designed for efficient operation in the temperature range of 200 to 300 deg C.

47. Holographic mirrors,
    Margarinos, J.R., Coleman, D.J., Optical Engineering
    Vol. 24 No. 5, (1985)

48. Holographic solar concentrators, A critical review,
    Ebbeni, J.P., SPIE Vol. 562, (1985)

49. Image collapsing concentrators,
    Sletten, C.J., DOE/CS/34163-2 (1980)

50. Linear Fresnel lenses for solar technology made of
    glass,
    Jirka, Vl., Maly, M., Nabelek, B., Triska, A., SPIE
    Vol. 502, (1984)

    Abstract:

    Linear Fresnel lenses (LFL) manufactured from glass are cheap and durable concentrators of direct solar radiation. Optical properties of LFL manufactured from glass are briefly discussed. Several examples of LFL applications are described.

51. Linear Fresnel lens with polar tracking,
    Kritchman, E.M., Applied Optics Vol. 20 no. 7, (1981)

    Abstract:

    The performance of coma and color corrected linear Fresnel lenses for solar concentration is evaluated for use in polar tracking systems. Effective concentrations of up to 90 at 75% efficiency were obtained.

52. Solar control tunable Lippmann holowindows,
    Jannson, J., Jannson, T., Yu, K.H., SPIE Vol. 562, (1985)

    Abstract:

    Tunable Lippmann Holowindows have been developed by NTS as a particular application for Broad-Band Lippmann Holograms investigated within Super High Effeciency (SHE) Holographic Technology. Those holograms can be tuned to be

highly reflective (above 99%) in any part of the near UV,
Visible and near IR. In particular, they can reflect all
direct solar heat radiation and additionally, the near UV
part of the solar spectrum that cannot be blocked by
glass or acrylic. Additionally, they can be combined with
conductive adhesives to create ideal cooling mirrors,
with parameters much better than those of existing products. Therefore, they can create a new generation of
high efficiency solar heat control windows whose
rejection of adjusted parts of the solar spectrum still
preserve high visible transmission (up to 87% for single
glazing), and the near perfect (close to 100%) reflection
of IR direct solar component. In the case of Tunable
Lippmann Holowindows, it is possible to adjust independently the U-value, emittance and shading coefficient by
the number of glazings, conductive adhesive thickness and
hologram reflection, respectively. In other words, SHE
Holographic Technology introduces to heat mirror
engineering a new degree of freedom - separate near IR
reflection coefficient.

53. Some advanced testing techniques for concentrator photo-
    voltaic cells and lenses,
    Wiczer, J.J., Chaffin, R.J., Hibray, R.E., SAND-82-2218C,
    (1982)

54. Testing of a prototype Fresnel-lens concentrator for
    thermal application,
    Lewandowski, A., Solar Engineering Vol. 5, (1984)

Abstract:

A prototype Fresnel lens concentrator, manufactured by
E-Systems, Dallas, Texas, was tested for thermal performance at SERI's Mid-Temperature Collector Research
Facility (MTCRF). This work was funded by the DOE in an
effort to support development of a testing standard for
concentrators. The data obtained from this testing was
presented and ultilized by a subcommittee of the American
Society for Testing and Materials (ASTM) in the development process for the standard. Several tests were conducted on the concentrator using draft versions of the
standard as guidance. Additional tests, allowed but not
reqired by the standard, were conducted to determine the
effect of the direct solar irradiance level on collector
performance. It is the results of these additional tests
that are of primary interest in this paper. The data
shows that non-linear heat losses cause collector effi-
cience to be a function of both t/IDN and IDN and that
the efficiency when fluid temperature is near ambient is
also a function of IDN. This latter result is a
characteristic unique to this collector, whereas the former holds for any collector with non-linear heat loss.

Stretched membrane

55. Analytical modeling and structural response of a
    streched-membran reflective module,
    Murphy, L.M., Sallis, D.V., SERI/TR-253-2101, (1984)

56. Derformation of polyester membranes with application to
    solar collectors,
    Authier B., Hill, L., Solar Energy Vol. 35, (1985)

Abstract:

A simple calculation method, validated by experimental
measurements, is proposed to determine the meridian line
of a pressurized, initially plane, elestic membrane. Provided the thickness variation is negligible, a nearly
spherical shape can be obtained when a horizontal disk is
filled with a liquid. Aluminized polyester film mirrors
so shaped are proposed to form solar concentrators. Dimensions of cost effective mirrors of this kind are therefore discussed as to physical feasability. Sphereshaped
polyester film mirrors are more suitable than inflated
ones for both parabolic dish and small fixed spherical
collectors.

57. Energy conversion system,
    Murphy, L.M., US-patent 6-776731, (1985)

58. Exposure testing of solar collector plastic films,
    Berry, M., Dursch, M., Solar Energy Materials 3, (1980)

59. Low-cost membrane solar concentrator demonstration,
    Jones, W.B., Clark, T., Wright, W., Solar Engineering
    Vol.8, (1986)

Abstract:

A significant portion of the cost of a solar thermal
power plant stems from the assemblies used to reflect and
concentrate the sunlight. Optical quality curved mirrors,
low-iron glass and heliostats have demonstrated high
reflective efficiencies but are not cost effective. More
recently, small membranes reflector/concentrator elements
(approximately 2 ft. in diameter) have been demonstrated
to be costeffective as components but the added cost of
necessary support structure and tracking systems are
still high. Using the unique fixed-mirror and distri-
buted-focus concept and adapting the structural fabric
design concepts, Texas Tech University plans to design,
build, and demonstrate a low-cost membrane solar concentrator. Design calculations, conducted using computerized
structural analysis methods, show structural requirements
for a 30 ft. diamter bowl can be met using fiber-rein-
forced plastics. Sub-scale tests are being used to evaluate membrane stiffness, have requirements, and construction assembly procedures. The 30 ft. diameter bowl will

be built and tested at the D.O.E. solar test site at Cros-
byton, Texas. The cost-effectiveness of the mebrane concentrator will be evaluated in terms of meeting local
mid-day peak demand electrical loads which are principal-
ly due to home air-conditioning.

60. Low-cost mirror concentrator based on inflated, double-
    walled, metallized, tubular films,
    Schwendemaŋ, J.L., Ball, G.L., Leffingwell, J.W.,
    McClung, C.E., Monsanto Reasearch Corp. TR/MRC-DA-944,
    (1981)

61. Membrane heliostat research,
    Murphy, L.M., SANDIA-85-8202, (1985)

    Abstract:

    An overview of the current status of the SERI research
    corresponding to the streched-membrane concept for helio-
    stat applications is presented. The focus is primarily on
    the reflector and the support structure (down to the
    drive attachment), which represents the largest fraction
    of the currently estimated total heliostat cost (about
    43%) and total weight (up to 85%, excluding the founda-
    tion). As such, the reflective module represents an impor-
    tant, but not the only, collector element that requires
    further development. The effect of greatly reduced weight
    in the reflector and support structure should have a
    positive cost impact on other elements. However, additio-
    nal development of drives, foundations, controls, and
    aerodynamic methods as well as other wind-avoidance
    schemes to reduce survival-level wind loading on the
    collector is both warranted and needed to meet installed
    heliostat field cost levels of $50-$60/m^2 and delivered
    energy costs of $5-$6/GJ.

62. Optical evaluation of cylindrical elastical concentrators
    McCormick, P.G., Solar Energy Vol. 26, (1981)

63. Polymer reflectors research during FY 1985,
    Schissel, P., SERI/PR-255-2835, (Forthcoming)

64. Single, stretched membrane, structural modules
    experiments
    Wood, R.L., SERI/TR-255-2736, (1986)

65. Solarkraftwerk mit einem Membranhohlspiegel 50 kW, Solar
    power-plant with a membrane concave mirror 50 kW,
    Benz, Bergermann, R., Schlaich und Partner Forschungsbe-
    richt 1986, (1986)

66. Specularity and stability of silvered polymers,
    Czandera, A.W., Schissel, P., SPIE Vol. 562, (1985)

67. Specular reflectance properties of silvered polymer
    materials,
    Susemihl, I., Schissel, P., SPIE Vol. 692, (1986)

Abstract:

The specular reflection properties of transparent cast
polymer sheets and extruded polymer films, silvered and
unsilvered, have been characterized with a newly designed
specular reflectometer. The results obtained with this
instrument are either absolute reflectances or a measure
for the Fourier transform of the reflection function of
the specimen in one dimension. Cast polymer sheets are
investigated before and after silvering, and silvered
polymer films are evaluated by mounting them with an
adhesive onto aluminum or glass substrates, or by suspend-
ing the thin, silvered polymer as a taut membrane. Silvered polymers have attained a specularity such that over
90% of the incident beam is contained in a 1-2 mrad
full-cone angle when mounted on a good substrate or suspended as a membrane. This value is well within the
current goals for solar concentrators but silvered polymer mirrors are currently less specular than glass
mirrors. The image quality of these mirrors does not
change significantly over the wavelength range 400 to
1000 nm and angles of incidence between 20 and 60 degrees. A Principal limiting factor to the initial specularity of the polymer mirrors was waviness and/or cur-
vature of the surface, hence, the material being used as
a substrate plays an important role in the optical performance of the mirror.

68. Stretched membrane heliostat research,
    Murphy, L.M., SERI/CP-251-2680, (1985)

69. Stretched membrane heliostats,
    N.N., Solar thermal technology, Annual evaluation report
    FY 1983 p. 73 ff, (1984)

70. System performance analysis of stretched membrane helio-
    stats,
    Anderson, J.V., Murphy, L.M., Short, W., Wendelin, T.,
    Solar Engineering Vol. 8, (1986)

71. System performance and cost sensitivity comparison of
    stretched membrane heliostat reflectors with current
    generation glass/metal concepts,Murphy, L.M.,
    SERI/TR-253-2694, (1985)

Abstract:

Heliostat costs have long been recognized as a major factor in the cost of solar central receiver plants. Research on stretched membrane heliostats has been emphasized because of their potential as a cost-effective
alternative to current glass/metal designs. However, the
cost and performance potential of stretched membrane

heliostats from a system perspective has not been studied until this time. The optical performance of individual heliostats is predicted here using results established in previous structural studies. These performance predictions are used to compare both focused and unfocused stretched membrane heliostats with state-of-the-art glass/metal heliostats from a system perspective. We investigated the sensitivity of the relative cost and performance of fields of heliostats to a large number of parameter variations, including system size, delivery temperature, heliostat module size, surface specularity, hemispherical reflectance, and macroscopic surface quality. The results indicate that focused stretched membrane systems should have comparable performance levels to those of current glass/metal heliostat systems. Further, because of their relatively lower cost, stretched membrane heliostats should provide an economically attractive alternative to current glass/metal heliostats over essentially the entire range of design parameters studied. Unfocused stretched membrane heliostats may also be attractive for a somewhat more limited range of applications, including the larger plant sizes and lower delivery temperatures.

72. Tensioning device for a stretched membrane collectur, Murphy, L.M., <u>US Patent PAT-APPL-440205,</u> (1982)

73. Technical and cost benefits of leigtweight, strefched-membrane heliostats, Murphy, L.M., <u>SERI/TR-253-1818,</u> (1983)

74. Technical and cost potential for leightweight, stretched membrane heliostat technology, Murphy, L.M., Yogi Goswami, D., <u>Solar Engineering,</u> (1984)

Abstract:

This paper presents the background and rationale and describes the development effort of a potentially low-cost, concentrating reflector design. The proposed reflector design is called the stretched-membrane concept. In this concept a reflector film-which can be metal, polymeric, or of a composite construction is stretched on a hollow torroidal frame that offers a structurally efficient and optically accurate surface. Although the intents is to improve heliostat concentrator cost and performance for solar thermal applications, the collector design approach proposed here may well offer effective cost and performance opportunities for improving photovoltaic and solar daylighting applications as well. Some of the major advantages include a reflector, a support frame, and support structures that can be made extremely lightweight and low in cost because of the effective use of material with high average stress levels

in the reflector and support frame; a 75% reduction in
the weight of the reflector and support structure (down
to the drive attachment) over the second-generation glass
and metal heliostat concept; a better than 50% cost reduction for the reflector assembly and support structure
compared to corresponding elements of the second-generation concept; and, finally, optical accuracies and an
annual energy delivery potential close to those
attainable with current glass-and-metal heliostats.

In this paper results are presented of initial design
studies, performance predictions, and analysis, as well
as results corresponding to subscale testing. Also included are recommendations for further development and
for resolving remaining issues.

75. The use of thin glass reflectors for solar concentrators,
    Marion, R.H., Solar Energy Materials 3, (1980)

Abstract:

Elastically deforming thin glass (thickness=0.13-0.80 mm)
provides an alternate method of formings a curved glass
reflector which can eliminate some of the disadvantages
of thicker glass. This paper describes a concept where
silvered thin glass is bonded to a steel backing to form
a laminate with a reflectance greater than 93%.
Subsequent bending of the flat reflector laminate to a
concentraiting profile produces compressive stresses
throughout the glass if the laminate is properly
designed. These compressive stresses enhance fracture
resistance and the lamination provides protection for the
silver. The design of the laminate is outlined for 0.25
and 0.51 mm thickness glass and fabrication procedures
are discussed. Thermal/humidity cycling hail impact, bond
strength measurements and reflectance results are
presented which demonstrate the performance capabilities
of this reflector laminate concept.

76. Variational approach for predicting the load deformation
    response of a double stretched membrane reflector module,
    Murphy, L.M., SERI/TR-253-2626, (1985)

Concentrator structures

77. A mini-size low cost heliostat system,
    Gerwin, H.L., Solar Energy Vol. 36 No. 1, (1986)

78. Analysis of static and quasi-static cross compound para-
    bolic concentrators,
    Molledo, A.G., Luque, A., Applied Optics Vol. 3, (1984)

79. Asymmetric second stage concentrators,
    Kritchman, E.M., Applied optics, 21 (1982)

    Abstract:

    Given a linear (2-D) primary optical concentrator and a
    convex close surface of any shape and location (within
    known limits) underneath, an ideal second-stage reflec-
    tive element that redirects the primary radiation onto a
    section of that surface may be designed. A useful appli-
    cation is to asymmetric two-stage concentrators for solar
    energy, the primary of which is not shaded from the
    illuminating source by the secondary.

80. Characteristics of the concentrated solar flux produced
    by the FMSC prototype,
    Harmon, S.Y., Backus, C.E., Pinon, R., Sharing the sun:
    solar technology in the seventies, Vol. 2, (1976)

81. Comment - the CPC concept - theory and practice,
    Tabor, H., Solar Energy 33, (1984)

82. Comparison of elliptical and parabolic non-imaging con-
    centrators,
    Gurnee, E.F., Solar Energy 19, (1977)

    Abstract:

    A recent paper has described the cross section of an
    ideal cylindrical concentrator as an elliptical segment,
    rather than the parabolic segment as described by
    Winston. A comparison is made between the two shapes, and
    it is shown that the elliptical cross section does not
    fulfill the requirements for an ideal cylindrical concen-
    trator.

83. Comparison of solar concentrators,
    Rabl, A., Solar Energy, 18 (1976)

84. Compound trapezoidal collector (an optimized
    stationary concentrator),

Villanueva, J., Truong, H.V., conference report of the "American section of the international solar energy society", Vol.1, (1977)

Abstract:

The design, evaluation, and optimization of a non-tracking compound trapezoidal groove collector (CTC) are presented. The proposed collector is compared to the Winston compound parabolic collector (CPC) and to the single trapezoidal groove collector. The proposed collector geometry consists of two successive trapezoidal grooves, the relative dimensions of which are optimized to accept (with no more than one reflection) all the solar energy impinging upon it when the sun's rays are directed along the optical axis of the collector. Computer simulation of the proposed collector shows that instantaneous concentration ratios as high as 5 (five) can be achieved with this geometry, while time concentrations of about 3.5 (during a complete collecting period of eight hours) are possible. The angular acceptance of this collector is quite reasonable and compares favorably with the truncated compound parabolic collector proposed by Winston and with trapezoidal groove collectors.

85. Cylindrical concentrators as a limit case of toroidal concentrators,
Minano, J.C., Applied Optics Vol. 23, No. 12, (1984)

86. Cylindrical parabolic solar mirrors,
Miyatani, K., Minematsu, K., Sato, I., Applied Optics Vol. 21 No. 24, (1982)

87. Deployment of a secondary concentrator to increase the intercept factor of a dish with large slope errors,
Ortabasi, U., Gray, E., O'Gallagher, J., Bericht der Queensland University N-84-28252, (1984)

Abstract:

The testing of a hyperbolic trumpet non-imaging secondary concentrator with a parabolic dish having slope errors of about 10 mrad is reported. The trumpet, which has a concentration ratio of 2.1, increased the flux through a 141-mm focal aperture by 72%, with an efficiency of 96%, thus demonstrating its potential for use in tandem with cheap dishes having relatively large slope errors.

88. Design and predicted performance of Scientific-Atlanta's fixed faceted mirror concentrator,
Shelton, S.V., Blackshaw, A., Hutchins, S.F., Solar concentrating collectors, (1977)

89. Design, construction and testing of a fixed mirror solar concentrator field,
Schuster, J.R., Russel, J.L., Eggers, G.H., Proceeding, Intersoc. Energy Convers. Eng. Conference Vol. 2, (1978)

90. Design of nonimaging concentrators as second stages in tandem with image-forming first-stage concentrators,
Winston, R., Welford, W.T., Applied Optics Vol. 19 No.3, (1980)

Abstract:

We show how paraboloidal mirrors of short focal ratio and similar systems can have their flux concentration enhanced to near the thermodynamic limit by the addition of nonimaging compound elliptical concentrators.

91. Design of two Receivers adapted to a fixed spherical solar collector,
Authier, B., Pouliquen, D., Solar Engineering, 4 (1981)

92. Design, performance investigation and delivery of a miniaturized Cassegrainian concentrator solar array,
Patterson, R.E., NASA-CR-178571, (1985)

93. Development of compound parabolic concentrators for solar energy,
O'Gallagher, J., Winston, R., Int. J. Ambient Energy Vol.4 (1983)

Abstract:

The compound parabolic concentrator (CPC) is not a specific collector, but a family of collectors based on a general design principle for maximizing the geometric concentration, C, for radiation within a given acceptance half angle = theta. This maximum limit exceeds by a factor of 2 to 4 that attainable by systems using focussing optics. The wide acceptance angles permitted using these techniques have several unique advantages for solar concentrators including the elimination of the diurnal tracking requirement at intermediate concentrations (up to about 10x), collection of circumsolar and some diffuse radiation and relaxed tolerances. Because of these advantages, CPC type concentrators have applications in solar energy wherever concentration is desired, e.g., for a wide variety of both thermal and photovoltaic uses. The basic priciples of nonimaging optical design are reviewed. Selected configurations for both non-evacuated and evacuated thermal collector applications are discussed with particular emphasis on the most recent advances. The use of CPC type elements as secondary concentrators is illustrated in the context of higher concentration photovoltaic applications.

94. Development of compound parabolic concentrators for solar
thermal applications
Allen, J., Levitz, N., Rabl, A., Reed, K., Schertz, W.,
Winston, R., American Society of Mechanical Engineering,
(1976)

95. Directional intercept factor of truncated CPC´s,
Minano, J.C., Applied Optics Vol. 22 No. 17, (1983)

96. Effect of diffuse radiation on a compound parabolic concentrator,
Yin, B.T., Bericht der "University of Singapore", (1979)

97. Effect of restricting the exit angle on the limit of concentration for cylindrical concentrators,
Minano, J.C., Applied optics, 23 (1984)

98. Fundamentals and techniques of nonimaging optics for solar enrgy concentration,
Winston, R., Report of the "University of Chicago",
(1980)

Abstract:

Nonimaging optics is a new discipline with techniques,
formalism and objectives quite distinct from the
traditional methods of focusing optics. These new systems
achieve or closely approach the maximum concentration
permitted by the Second Law of Thermodynamics for a given
angular acceptance and are often called ideal. Applaca-
tion of these new principles to solar energy over the
past seven years has led to the invention of a new class
of solar concentrators, the most well known version of
which is the Compound Parabolic Concentrator or CPC. A
new formalism for analyzing nonimaging systems in terms
of a quantity called the geometrical vector flux has been
developed. This has led not only to a better under-
standing of the properties of ideal concentrators but to
the discovery of several new concentrator designs. One of
these new designs referred to as the trumpet concentrator
has several advantageous features when used as a
secondary concentrator for a point focusing dish
concentrator. A new concentrator solution for absorbers
which must be separated from the reflector by a gap has
been invented. The properties of a variety of new and
previously known nonimaging optical configurations have
been investigated: for example, Compound Elliptical
Concentrators (CEC´s) as secondary concentrators and
asymmetric ideal concentrators. A thermodynamic model
which explains quantitatively the enhancement of effective absorptance of gray body receivers through cavity
effects has been developed. The classic method of Liu and
Jordan, which allows one to predict the diffuse sunlight
levels through correlation with the total and direct
fraction was revised and updated and applied to predict
the performance of nonimaging solar collectors. The conceptual design for an optimized solar collector which
integrates the techniques of nonimaging concentration
with evacuated tube collector technology was carried out.

99. High temperature solar collector of optimal cuncen-
    tration: non-focusing lens with secondary concentrator,
    Collares-Pereira, M., O´Gallagher, J., Rabl, A., Winston,
    R., Egger, J., Williams, K., Sun: mankinds future source
    of energy Vol. 2

100. Generalized conic concentrators,
     Eichhorn, W.L., Applied Optics Vol. 21, No. 21, (1982)

101. Geometric characteristics of ideal nonimaging (CPC)
     solar collectors with cylindrical absorber,
     Baum, H.P., Gordon, J.M., Solar Energy Vol. 33, (1984)

     Abstract:

     Analytic expressions are derived for the collector height
     and reflector arc length of an ideal nonimaging collector
     with a cylindrical absorber, for an arbitrary degree of
     reflector truncation, which will serve as a major step in
     the design of optimized collectors. It is noted that,
     while there is complete agreement between the analytic
     result and the numerical results of McIntire (1979) for
     collector height, significant differences emerge in the
     results for reflector arc length. 7 references.

102. High concentration solar collector of the stepped spheri-
     cal type: optical design characteristics,
     Authier, B., Hill, L., Applied Optics Vol. 19, No. 20,
     (1980)

103. High temperature solar concentrator design,
     Wientzen, R., Davis, W.J., Forky, R.E., Proc. Soc. of
     Photo-Opt. Instru. and Engineering Vol. 237, (1980)

104. Ideal second stages in tandem with parabolic concentra-
     tors,
     Kritchman, M.E., Applied optics, 21 (1982)

     Abstract:

     Expressions for the concentration capability of parabolic
     reflectors are derived and extended to include additional
     ideal second-stage concentration. The ultimate concentra-
     tion of a parabolic reflector (primary) alone is shown to
     achieve one-half of the ideal limit in 2-D geometry; the
     respective two-stage configuration is shown to approach
     the ideal limit itself at large f/ratio.

105. Ideal second stages in tandem with spherical mirrors,
     Kritchman, E.M., Solar Energy Vol. 30, (1983)

     Abstract:

     Even though, as a means to concentrate solar energy, a

spherical mirror is less concentrated than a parabolic mirror, the spherical mirror may be emphasized because of its low cost. The concentration of the spherical mirror can also be improved by the introduction of second stage elements at their focal plane. Cylindrical mirror geometry is outlined. Concentration factors are derived by means of equations. It is determined that an ideal concentration element may be installed at the focal neighborhood of the primary to redirect the rays into a small exit aperture, and increase concentration. The formula for the combined two stage concentration is derived, Finally, this mirror is compared to the parabolic mirror.

106. Innovative solar thermal dish technology development, Schwinkendorf, W.E., SAND-84-7011, (1984)

Abstract:

A Cassegrainian point focus solar concentrator system has been analyzed and a conceptual design developed. In this system, the receiver is located at the vertex of the primary mirror eliminating limitations of receiver size and weight associated with standard parabolic dish collectors. Disadvantages include increased reflection loss at the secondary mirror and increased beam spread associated with a longer focal length. A non-imaging, trumpet shaped tertiara reflector located at the receiver aperture increases the system efficiency by 15 to 20%. Because the secondary mirror may reach very high temperatures, a reflective film cannot be used as a mirror surface. Recommended instead is thin polished stainless steel stamped into shaped and used as the mirror substrate with vacuum deposited silver used for the reflecting surface. A transparent coating is then required to protect the silver. Both the secondary and primary mirrors require radial supports on the back surface to minimize deflection under wind loading. The Cassegrainian system is more efficient than the standard dish only at high operation temperatures and for large receivers. However, the cost per kilowatt into the receiver aperture is less for the Cassegrainian, due primarily to the high cost of piping and insulation running to the receiver of the standard dish. Further efforts in the analysis and design of the Cassegrainian concept appear warranted.

107. Intensity distribution in cylindrical-circular receivers for nonperfect cylindrical-parabolic concentrators, Nicolas, R.O., Applied Optics Vol. 24, No. 16, (1985)

108. Limit of concentration for cylindrical concentrators under extended light sources, Minano, J.C., Luque, A., Applied Optics Vol.22, No.16, (1983)

109. Maximally concentrating collectors for solar energy applications, Guay, E.J., Solar Energy 24, (1980)

110. Minimum-mirror-area single-stage solar concentrators
     Mills, D., Harting, E., Giutronich, J.E., Cellich, W.,
     Morton, A., Walker, I., Optical Society of America (1980)

111. New concentrators for the generation of very high tempe-
     ratures from solar energy,
     Winston, R., O'Gallagher, J.O., DOE Final technical
     report, Grant No. DE FG02-79ET-00089, (1983)

112. New family of 2-D nonimaging concentrators: the compound
     triangular concentrators,
     Minano, J.C., Applied optics, 24 (1985)

     Abstract:

     A new family of 2-D nonimaging concentrators is
     presented. The most significant characteristics of these
     concentrators are their small size and the fact that they
     use straight, as opposed to curved, reflective, or
     refractive interfaces. The concentrators are filled with
     a medium of refractive interfaces. The concentrators are
     filled with a medium of refractive index n>1 and use
     narrow strips of refractive index n<n. These strips are
     transparent, or act as mirrors, depending on the angle of
     incidence of the rays. The direction of the collected
     rays (i.e., rays reaching the receiver) at their entry
     aperture can be varied by modifying the refractive
     indices of the strips.

113. Non-focussing solar concentrators of easy manufacture,
     Shapiro, M.M., Solar Energy Vol. 19 No. 2, (1977)

114. Nonimaging secondary concentrators,
     Winston, R., O'Gallagher, J., Jet Propulsion Lab.
     JPL-PUB-83-2, (1983)

     Abstract:

     Secondary concentrators deployed at the focal plane of a
     parabolic dish can significantly increase the system
     concentration ratio or alternatively decrease the to-
     lerance requirement. Several trumpet shaped radiation
     flow line concentrators were tested with the JPL Test Bed
     Concentrator at the Parabolic Dish Test Site in the
     Mojave Desert. Primary flux inside an 8 inch diameter
     circle was redirected into 5 1/2 inches with an
     efficiency exceeding 96%. A power gain of 30% was
     observed.

115. Nonimaging second-stage elements: a brief comparison,
     Kritchman, E.M., Applied Optics, 20 (1981)

116. Operating experience with the General Atomic fixed mirror solar concentrator,
Schuster, J.R., Eggers, G.H., Russell, J.L., Solar diversification Vol. 2.1, (1978)

117. Optical analysis of Cassegrainian concentrator systems,
Bass, A.H. Jr., Schrenk, G.L., Poon, P.T.Y., Higgins, S.N., SUN II, Conference report CONF-790541-Vol.2, (1979)

118. Optical analysis of paraboloidal concentrators,
McDanal, A.J., SUN II, Conference report CONF-790541-Vol.2 (1979)

119. Optical and thermal analysis of concentrators,
Rabl, A., SERI/TR-34-048, (1978)

120. Optical and thermal design considerations for ideal light collectors,
Goodman, N.B., Rabl, A., Winston, R., Conference report "Sharing the sun" Vol. 2 CONF-760842-P2, (1976)

121. Optical and thermal properties of compound parabolic concentrators,
Rabl, A., Solar energy, 18 (1976)

Abstract:

Compound Parabolic Concentrators (CPC) are relevant for solar energy collection because they achieve the highest possible concentration for any acceptance angle (tracking requirement). The convective and radiative heat transfers through a CPC are calculated, and formulas for evaluating the performance of solar collectors based on the CPC principle are presented. A simple analytic technique for calculating the average number of reflections for radiation passing through a CPC is developed; this is useful for computing optical losses. In most practical applications, a CPC will be truncated because a large portion of the reflector area can be eliminated without seriously reducing the concentration. The effects of this truncation are described explicitly. The paper includes many numerical examples, displayed in tables and graphs, which should be helpful in designing CPC solar collectors.

122. Optical properties of compound circular arc concentrators,
Jones, R.E., Anderson, G.C., Solar Energy 21, (1978)

123. Optical properties of compound elastical concentrators,
McCormick, P.G., Solar Energy Vol.30 No.6, (1983)

Abstract:

In summary, it has been shown that simple elastic bending
can be used to form the reflector profile for a
non-imaging concentrator. For low concentration with flat
or cavity absorbers approach that of truncated CPC
concentrators.

124. Optics of nonimaging concentrators. Light and solar
     energy
     Welford, W.T., Winston, R. Book, Academic Press Incorpo-
     rated, NY, (1978)

125. Optimization of dish solar collectors,
     Jaffe, L.D., Journal of Energy Vol.7 No.6, (1983)

126. Optimisazation of dish solar collectors with and without
     secondary concentrators,
     Jaffe, L.D., JPL-PUB-82-103, (1982)

127. Optimized second stage concentrator,
     Kritchman, E., Applied optics, 20 (1981)

128. Planar-sectioned solar concentrators. 1.: Polygon re-
     flectors
     Hamadto, S.A., Applied Optics Vol.23, No.8, (1984)

129. Point focus dishes with relaxed otipcal tolerances ulti-
     lizing non-imaging secondary concentrations,

130. Preliminary testing of a scale model secondary concen-
     trator for the Sandia Solar Thermal Test Facility,
     Mulholland, G.P., Matthews, L.K., Conference report Nr.
     CONF- 790445, (1979)

131. Principles of fixed mirror solar concentrator,
     Russel, J.L., Optics in solar energy utilization II,
     (1977)

132. Rapid test bed concentrator (TBC) alignment techniques,
     Argoud, M.J., TR/Jet Propulsion Lab. N-84-28256, (1984)

133. Secondary and compound concentrators for parabolic-dish
     solar-thermal power systems,
     Jaffe, L.D., Poon, P.T., Jet Propulsion Lab. JPL-PUBL-
     8127 (1981)

134. Secondary concentration in central receivers. Application
     to the volumetric receiver,
     Blanco, M., Carmona, R., Martin, J.G., Institute of rene-
     wable energies- JEN, Madrid, Spain, (1986)

Abstract:

This .letter expresses the interest of the Institute of Renewable Energies (Spain) and faculty of the University of Lowell (USA) to participate in second stage concentration in central solar receivers. The goal of the proposal is to design, construct, operate, and test an actual full-scale secondary concentrator at the IEA/SSPS facility. It aims at tapping a considerable level of interest in high concentration in the international community, so as to make an effective contribution to the state of the art. The installation of the terminal concentrator to the volumetric reciver should enable the evaluation of the reconcentrator and the enhancement of the volumetric receiver performance.

The project consists of three phases-design, construction, and testing. In the design-phase, suggested configurations (i.e., compound elliptic, compound parabolic, trumpet, etc.) will be surveyed and their expected performance simulated utilizing actual plant data. Materials will be evaluated with evaluated with respect to performance requirements. Special emphasis will be made on ceramics and composite materials to minimize stress problems and optical and structural degradation. The product of this phase will be a working design of a secondary concentrator.

Actual construction of the concentrator will be the task in the second phase, scheduled for the second year. Construction at the plant, operation and testing for a period of about three months, will constitute the final phase.

135. Terminal concentrators (An analytical examination) Sizmann, R., theoretical survey (1984)

136. Testing the figure of parabolic reflectors for solar concentrators, Bodenheimer, J.S., Eisenberg, N.P., Gur, J., Applied Optics Vol.21, No.24, (1982)

137. Tower reflector for solar power plant (technical note) Rabl, A., Solar energy, 18 (1976)

138. Truncated compound parabolic concentrator analysis, Almonacid, G., et al, MELECON '85, Vol.IV, Elsevier, Holland, (1985)

139. Truncation of CPC solar collectors and it's effect on energy collection,

Carvalho, M.J., Collares-Pereira, M., Gordon, J.M., Rabl,
A., Solar Energy 35, (1985)

Abstract:

Analytic expressions are derived for the angular
acceptance function of two-dimensional compound parabolic
concentrator solar collectors (CPC's) of arbitrary degree
of truncation. Taking into account the effect truncation
on both optical and thermal losses in real collectors, we
also evaluate the increase in monthly and yearly
collectible energy. Prior analysis that have ignored the
correct behavior of the angular acceptance function at
large angles for truncated collectors are shown to be in
error by 0-2% in calculations of yearly collectible
energy for stationary collectors.

140. Adhesion and hardness of ion-plated TiC and TiN coatings,
Hummer, E., Perry, A.J., Thin Solid Films 101 (1983)

Abstract:

The microhardness and adhesion of TiC coatings on steel
and cemented carbide are studied. A comparison with TiN
coatings is also made. It is found that the minimum
coating thickness in which a reliable microhardness
measurement can be made is described satisfactorily by
existing standards; departures can be expected if the
coating is strongly textured. The critical load for the
removal of the coating by the scratch test depends on
both the coating and the substrate and is thought to
reflect their mechanical properties. It appears that
marked variations in the strength and nature of adhesion
on cemented carbides can occur.

141. Characterization of titan nitride films deposited onto
silicon,
Armigliato, A., Celotti, G., Garulli, A., Guerri, S.,
Ostoja, P., Rosa, R., Martinelli, G., Thin Solid Films,
92 (1982)

Abstract:

Films of titanium nitride were prepared by reactive
evaporation and r.f. sputtering and were characterized
from the optical, electrical, chemical and structural
points of view. The effect of oxygen, introduced during
the deposition process as well as in a subsequent thermal
annealing, on the properties of the films is reported.
The applicability of titanium nitride to silicon solar
cells as a transparent conducting material is briefly
discussed.

142. Comparison of the properties of ion-plated TiC films prepared by different activation methods,
Fukutomi, M., Fujitsuka, M., Okada, M., Thin Solid Films,
120, (1984)

A comparative study was made of titanium carbide films
ion plated onto molybdenum by d.c. and r.f. discharge
methods, from the point of view of the characteristics of
the discharge and the quality of deposits made by each
technique. A very high specimen current was achieved in
the d.c. discharge method. Stoichiometric TiC deposits
were easily obtained in a wide range of the pressure

ratio of C_2H_2 gas to titanium vapour. In contrast, careful optimization of the deposition conditions was needed in the r.f. discharge method for controlling the stoichiomety of the deposits. Titanium carbide coatings deposited onto molybdenum by the d.c. discharge method showed excellent thermal stability compared with those prepared by the r.f. discharge method. These results indicate the strong influence of the ionization efficiency of the properties of ion-plated titanium carbide deposits.

143. Composition, optical properties and degradation modes of Cu (Graded metal-carbon) solar selective surfaces, Craig, S., Harding, G.L., Thin Solid Films, 101 (1983)

Abstract:

A detailed study is reported of the composition and properties of a production Cu/(graded stainless steel-carbon) selective surface which is incorporated in all-glass evacuated solar thermal collectors. The Auger depth profile, reflectance for wavelengths in the range 0,35 - 2,5 um and temperature-dependent emittance for the range 100 - 300 Grad C were determined for the selective surface before and after heat treatment of the collector at 500 Grad C. The composition, carbon atom bonding states, electrical resistivity, refractive indices (wavelengths, 0,35 - 2,5 um), IR transmittance (Wavelengths, 2.5 - 16 um) and emittance (100 - 300 degrees C) for a set of homogeneous stainless steel-carbon component layers of the graded layer profile were also determined before and after heat treatment. In addition, a multi-layer stack computer model based on the composition profile of the production selective surface and the comprehensive refractive index data for the set of homogeneous component layers was developed. Correlation of the experimental data for the set of homogeneous layers with alterations in reflectance for the multilayer stack resulting from the substitution or removal of component layers in the stack allowed a detailed evaluation of the factors determining production selective surface properties before and after heat treatment. A theoretical study of alternative grading profiles with a view to optimizing the solar selective properties of the surface is also discussed.

144. Contact resistiveties of sputtered TiN and Ti-TiN metallisations on solar-cell-Type silicon,, Mäenpää, M., Nicolet, M.A., Suni, I. Solar energy 27 (1981)

145. Corrosion behaviour and protective quality of TiN
     coatings,
     Mätylä, T.A., Helevirta, P.J., Lepistö, T.T., Siitonen,
     P.T., Thin Solid Films 126 (1985)

Abstract:

TiN hard coatings made by different chemical and physical
vapour deposition techniques are widely used to increase
the wear resistance of steel components. In many applications there are also corrosive stresses combined with
different wear mechanisms, but so far little information
is   avaiable on the corrosion behaviour of TiN coatings.

In the study we determined the corrosion behaviour of
various chemically vapour-deposited TiN coatings   by
electrochemical polarization measurements in an aerated
0,01 N HCl aqueous solution. The effect of some process
parameters, e.g. temperature, TiCl4 concentration and gas
flow rate, on the protective quality of the coatings was
also studied. The polarization of the TiN coatings was
compared with the behaviour of pure TiN and with that of
the substrates. The observed polarization behaviour of
TiN coatings was correlated to structural details studied
by scanning electron microscopy. The results of these
tests are presented and discussed.

The polarization measurements proved to be very sensitive
to different structural defects and this type of measurement can be used to determine the protective quality of
thin coatings of the type studied.

146. Dekorative Beschichtungen von keramischen Erzeugnissen
     auf der Basis von gold- und platinfarbenen Titannitrid-
     schichten im Hochvakuum,
     Junghans, W., Wilberg, R., Liebmann, H., Schmidt, H-J.,
     Silikattechnik 33 (1982) Heft 12

147. Deposition of hard wear-resistant coatings by reactive
     d.c. plasmatron sputtering,
     Schiller, S., Heisig, U., Beister, G., Steinfeldar, C.,
     Strämpfel, J., Korndörfer, Chr., Sieber, W., Thin Solid
     Films, 118 (1984)

148. Doped TiO for solar energy applications,
     Wong, W.K., Malati, M.A., Solar energy 36 (1986)

149. Effects of substrate temperature and substrate materials
     on the structure of reactively sputtered TiN films,
     Hibbs, M.K., Johansson, B.O., Sundgren, J.-E.,
     Helmersson, U., Thin Solid Films, 122 (1984)

Abstract:

Stoichiometric TiN films were reactively magnetron sputtered in an Ar-N atmosphere. The films were deposited at
various substrate temperatures in the range of 200 - 650
degrees C two types of substrate material, high speed
steel and stainless steel. The microstructure of the films
obtained was investigated by the use of a transmission
electron microscope and the morphology was studied in a
scanning electron microscope. Measurements of the hardness were also performed. The analysis of the microstructure shows that the growth of the film is markedly
influenced by the substrate material. In particular, the
high speed steel substrates were found to have a considerable influence on the microstructure. The vanadium carbide
particles in these steels, which have a good lattice match
to TiN, stimulate a localized epitaxial growth to occur on
these carbide particles. This results in a microstructure
consisting of large grains surrounded by small grains. The
shape of the large grains is influenced by the temperature. In the development of these large grains cracks
and/or voids occur in and around the grains at substrate
temperatures above 400 degrees C and the hardness drops by
about 20%. No large grains were found on films deposited
onto stainless steel and their hardness increases slightly
with temperature. High hardness for films deposited onto
the high speed steel substrates at temperatures above 400
degrees C can also be obtained if a substrate bias is
used. Ion bombardement during film growth suppresses the
formation of the large grains witch voided or cracked
boundaries because of a continuous renucleation process.
The formation of the different microstructures is discussed in terms of surface energy minimization and
thermally activated processes as surface and grain
boundary migration.

150. Effects of the experimental conditions of chemical vapour
     depositions on a TiC/TiN double-layer coating,
     Kim, M.S., Chun, J.S., Thin Solid Films 107 (1983)

151. Fabrication and testing of trilayers with a high
     deposition rate plasma-polymerized spacer,
     Mazzeo, N.J., Ahn, K.Y., Jipson, V.B., Lynt, H.N., Thin
     Solid Films, 108 (1983)

152. Influence of the growth conditions on the optical
     properties of thin gold films,
     Parmigiani, F., Scagliotti, M., Samoggia, ff, Ferraris,
     C.P., Thin Solid Films, 125 (1985)

153. Influence of the nitrogen partial pressure on the
     properties of d.c. sputterd Ti and TiN-films,
     Lemperiere, G., Poitevin, J.M., Thin Solid Films, 111
     (1984)

Abstract:

The properties of titanium and titanium nitride films
deposited onto biased substrates in a d.c. sputtering
system were studied as a function of the partial nitrogen
pressure. The deposition rate was deduced from film
thickness measurements. The film composition was determined by Rutherford backscattering analysis and the structure was studied using X-ray diffraction. The resistivity
was measured by the four-probe method and the temperature
coefficient of resistivity (TCR) was determined in the
temperatur range from -196 to 25 degrees centigrade.

Around a critical nitrogen pressure p of 4×10^{2}- Pa
the deposition rate decreases rapidly, the film structure
changes from h.c.p. titanium to nearly stoichiometric
f.c.c. TiN. At the same pressure, the film resistivity
and the TCR present minimum values.

A general sputtering model which takes into account the
gettering action of the deposited material is proposed.
This model allows the calculation of the surface coverage
of the target by the reactive gas or the metallic compound
and the determination of the deposition rate as a
function of the reactive partial pressure. A good
agreement is found with the deposition rates measured
experimentally.

154. Investigation of TiN films reactively sputtered using a
     sputter gun,
     Ahn, K.Y., Wittmer, M., Ting, C.Y., Thin Solid Films 107
     (1983)

155. Material selection for hard coatings,
     Holleck, H., Int. conf. on metallurgical coatings, San
     Diego, (1986)

156. Materials for solar-transmitting heat-reflecting
     coatings,
     Karlsson, B., Valkoen, E., Karlsson, T., Ribbing, C.G.,
     Thin Solid Films 90 (1982)

157. Materials for solar-transmitting heat-reflecting
     coatings,
     Karlsson, B., Valkonen, E., Karlsson, T., Ribbing, C.-G.,
     Thin Solid Films 86 (1981)

Abstract:

A coating for solar energy applications which combines heat reflection with transparency to solar radiation may bo of four different types: a metallic film which is ufficiently thin to be transparent; a metal-based multi-layer coating; a wide band gap heavily doped semiconductor such als SnO2 ot In2O3; a conducting microgrid. We prepared such coatings on glass by evaporating thin films of silver, copper, gold, aluminium, cobalt, iron, chromium and nickel of various thickness and by spraying SnO2 films. The spectral variations in the transmittance, and the front side and back side reflectances were measured in the wavelenght range 0,4 - 15 um. The properties of a three-layer coating of the dielectric/-metal/dielectric type were calculated with a multilayer programm using known bulk optical constants. The effect ot these films when coated onto a domestic windows was demonstrated with a heat transfer calculation using an equivalent thermal net. When a large transmittance over a broad range of the solar spectrum is required, gold is an equally good, or a slightly better, choice than silver as the metal in a three-layer coating. In general, an SnO2 film exhibits a higher solar transmittance as well as a higher emittance than a coating containing metals. This implies that the oxide is to be preferred as a coating on a window when the maximum passive solar heating is sought. However, a metalbased coating could be better when a very low U_L value is the most important require-ment.

158. Mechanical properties of sputtered TiN coatings,
Milic, M., Milosavljevic, M., Bibic, N., Nenadovic, T.,
Thin Solid Films, 126 (1985)

159. Mechanisms of reactive sputtering of titaniumnitride and titaniumcarbide,
1. Influence of process parameters on filmcomposition,
Sundgren, J.-E., Johannson, B.O., Karlsson S.-E., Thin Solid Films 105 (1983)
2. Morphology and structure, Sundgren, J.-E., Johansson, S.-O., Karlsson, S.-E., Hentzell, H.T.G., ebenda
3. Influence of substrate bias on composition and struc-ture,
Sundgren, J.-E., Johansson, B.-O., Hentzell, H.T.G., Karlsruhe, S.-E., ebenda

160. Mechanisms of the bias sputtering of titanium in an Ar-N2 mixture,
Poitevin, j.M., Lemperiere, G., Thin Solid Films, 120 (1984)

161. Microstructure and hardness Ti(C,N) coatings on steel
     prepared by the activated reactive evaporation technique,
     Jacobson, B.E., Deshpandey, C.V., Doerr, H.J., Karim,
     A.A., Bunshah, R.F., Thin Solid Films, 118 (1984)

162. Microstructure and mechanical properties of TiC-Al2O3
     coatings,
     Budhani, R.C., Memarian, H., Doerr, H.J., Deshpandey,
     C.V., Bunshah, R.F. Thin Solid Films, 118 (1984)

163. Moderne Oberflächen- und Dünnschichtentechnologien, Ver-
     fahren und Anwendungen,
     Münz, W.D., VDI-Tagung 22. April 1986, (1986)

164. Morphology and properties of sputtered TiN layers as a
     functuion of substrate temperature and sputtering
     pressure,
     Kopacz, U., Jehn, A., Thin Solid Films, 126 (1985)

165. Morphology and structure of ion-plated TiN, TiC and
     Ti (C,N) coating,
     Gabriel, H.M., Kloss, K.H., Thin Solid Films, 118 (1984)

166. Nobel-metal-based transparent infrared-reflectors: Pre-
     paration and analysis of thin gold films,
     Smith, G.B., Niklasson, G.A., Svensson, J.S.E.M., Gran-
     qvist, C.G., SPIE Vol. 562, (1985)

167. Optical and electrical properties of reactively d.c.
     magnetron-sputtered In2O3: Sn films,
     Theuwissen, A.J.P., Declerck, G.J., Thin Solid Films, 121
     (1984)

168. Optical Properties of CVD-Coated TiN, ZrN and HfN,
     Karlsson, B., Shimhock, R.P., Seraphin, B.O., Haygarth,
     J.C., Physica Scripta Vol. 25, (1982)

169. Optical Properties of CVD-Coated TiN, ZrN and HfN,
     Karlsson, B., Shimshock, R.P., Seraphin, B.O., Haygath,
     J.C., Solar Energy Materials 7, (1983)

170. Optical properties of SiO2-TiO2-composite films,
     Demiryont, H., Applied Optics Vol. 24, No. 16 (1985)

171. Optical properties of transparent heat mirrors based on
     thin films of TiN, ZrN, HfN,
     Karlsson, B., Ribbing, C.G., SPIE Vol. 324 (1982)

Abstract:

Calculations of the transmittance and reflectance between
0,35 μm and 10 μm of semitransparent films of TiN, ZrN
and HfN have been performed. The calculations are based

on recently reported optical constants. They show that
these compounds can be used as transparent heat-mirrors.
These materials show considerable higher emittance than
the noble-metals but comparable or higher visible transmittance. It is also shown that the transmittance can be
increased by the technique of induced transmission.

172. Optical scattering and absorption losses at interfaces
and in thin films,
Bennett, J.M., Thin Solid Films, 123 (1985)

173. Optimized transparent and heat reflecting oxide and
nitride films,
Howson, R.P., Ridge, M.I., Suzuki, K., SPIE Vol. 324
(1982)

174. Oxidation kinetics of ZrN thin films,
Krusin-Elbaum, L., Wittmer, M., Thin Solid Films 107
(1983)

175. Photoelectrochemical properties of plasma-deposited
TiO_2 thin films,
Williams, L.M., Hess, D.W., Thin Solid Films, 115 (1984)

176. Preparation and characterization of carbon and titanium
carbide coatings,
Dua, A.K., George, V.C., Agarwala, r.P., Krishnan, R.,
Thin Solid Films, 121 (1984)

177. Preparation, characterization and stablility tests of
selectively absorbing oxides on stainless steel,
Karlsson, R., Valkonen, E., Solar Energy Vol. 34 No.4/5,
(1985)

Abstract:

Selectively absorbing surfaces have been prepared on
ferritic stainless steel by thermal oxidation and four
different wet chemical methods: sodium dichromate.
Ebonol, chemical coloured and chemical coloured followed
by cathodic hardening. The solar optical parameters were
determined by spectral reflectance measurements and
values obtained were around 0,90 and e_{373} values in the
interval 0,10 - 0,15. The surface oxides have been
analyzed by x-ray diffraction, ESCA and AES. The Ebonol
treatment resulted in $Fe3O4$ while the other methods
produced surfaces mainly consisting of chromium oxide
with some content of iron oxide. The selective surfaces
have been exposed to accelerated and natural aging.

Optical and surface analysis did not reveal any degradation of the samples. The surfaces oxidized by sodium
dichromate and in particular the thermally oxidized
surfaces exhibited excellent temperature stability. In
low temperature applications the chemically coloured
surface is superior owing to its better solar selectivity.

178. Problems in the physical vapour deposition of titaniumnitride (TiN),
Matthews, W., Lefkow, A.R., Thin Solid Films, 126 (1985)

179. Properties of TiNx films reactively sputtered in an argon
nitrogen atmosphere,
Posadowski, W., Krol-Stepniewska, L., Ziolowski, Z.,
Thin Solid Films, 62 (1979)

180. Properties, applications and manufacture of wearresistant
hard material coating for tools,
Schintlmeister, W., Wallgram, W., Kanz, J., Thin Solid
Films 107 (1983)

181. r.f. reactively sputtered TiN: characterization and
adhesion to materials of technical interest,
Bucher, J.P., Ackermann, K.P., Buschor, F.W., Thin Solid
Films, 122 (1984)

182. Reactive sputtering of titanium and properties of
titanium suboxide films for photochemical applications,
Soliman, A.A., Thin Solid Films 100 (1983)

183. Reactive sputtering of nitrides and carbides,
Münz, W.D., Special Application 11-S07.2, Leybold-
Heraeus, (1985)

184. Reflectivity of ScNx thin films: Comparsion with TiNx,
TiNxCy and ZrNx coatings and application to the photo-
thermal conversion of solar energy,
Francois, J.C., Chassaing, G., Pierrisnard, R., Bonnot,
A.M., Thin Solid Films, 127 (1985)

Abstract:

This study is concerned with the room temperature reflectivity of sputtered opaque films of scandium, titanium
and zirconium nitrides and carbonitrides in the energy
range 0,22 - 5,50 e V. The reflectivity curves show two
frequency regions, an absorbing region in the visible and
near-IR range and a reflecting region in the IR range,
which are clearly distinguished and separated at a
critical energy. The cut-off energy depends on the nature
and concentration of the components and is related to the

number of free electrons at the Fermi level. If x is the atomic ratio, of nitrogen to titanium we show the cut-off is red shifted when x increases or when nitrogen is replaced by carbon. When titanium is replaced by zirconium we observe the same behavior. The results for the optical properties of ScNx films with low nitrogen concentrations x are the best we have ever obtained for solar energy conversion.

185. Residual compressive stress in sputter-deposited TiC-films on steel substrates,
Panm, A., Greene, J.E., Thin Solid Films, 78 (1981)

186. Sachverzeichnis: Silikattechnik ('82)

187. Scratch adhesion testing of hardcoatings,
Perry, A.J., Thin Solid Films, 107 (1983)

188. Selective transmission of thin TiN-films,
Valkonen, E., Karlsson, T., Karlsson, B., Johansson, B.O., Thin Film Technolgies, Proc. Soc. Photo-Opt. Instrum. Eng. (401), (1983)

Abstract:

The solar selective properties of thin titanium nitride films have been studied. High quality TiN-films were made on fused silica and glass substrates by reactive sputtering of titanium in a nitrogen atmosphere. The resulting film thicknesses ranged from 5 - 120 nm. The reflectance and transmittance measurements confirm the high solar and visible transmission previously calculated from optical constants. The infrared reflection is lower than calculated, but sufficiently high to make TiN a new competitor for selective transmission applications. Three optical parameters were determined and used to obtain the optical constants as well as the film thickness. The refractive index increases with decreasing thickness. The extinction coefficient is almost constant above the thickness of 12 nm, but strongly reduced for films thinner than this.

189. Solar selective titanium oxinitride films prepared by reactive sputtering,
Vogelzang, E., Sikkens, M., SPIE Vol. 502 (1984)

190. Sputtered In2O3: Sn films: Preparation and optical properties,
Jiang, S.-J., Granqvist, C.G., SPIE Vol. 562, (1985)

191. Sputtering system for multicharacteristic solar window films,
Meckel, B.B., Thin Solid Films, 108 (1983)

192. Structure and properties of TiN coatings,
Sundgren, J.-E., Thin Solid Films, 128 (1985)

193. The adhesion between physically vapour-deposited or chemically-vd alumina and TiC coated cemented carbides as characterized by auger electron spectroscopy and scretch testing,
Lhermitte-Serbire, I., Colmet, R., Naslain, R., Desmaison, J., Gladel, G., Thin Solid Films, 138 (1986)

194. The interpretation of weak inhomogeneities in TiO_2 layers,
Jankuj, J., Thin Solid Films, 116 (1984)

195. The microstructure of reactively sputtered TiN films,
Hibbs, M.K., Sundgren, J.-E., Jakobson, B.E., Johansson, B.-O., Thin Solid Films, 07 (1983)

196. The Optical Properties of Titanium Nitrides and Carbides: Spectral Selectivity and Photothermal Conversion of Solar Energy,
Roux, L., Hanus, J., Francois, J.C., Sigrist, M., Solar Energy Materials 7, (1982)

197. Theoretical model for the optical properties of In2O3: Sn films in the 0,3 - 50 um range,
Hamberg, I., Granqvist, C.G., SPIE Vol. 562, (1985)

198. The stress in ion-plated HfN and TiN coatings,
Chollet, L., Perry, A.J., Thin Solid Films, 123 (1985)

199. Ti (C,N,H) coatings on glass substrates prepared by chemical vapour deposition using Tris (2,2`-bipyridine) titanium (O), Morancho, R., Constant, G., Ehrhardt, J.J., Thin Solid Films, 77 (1981)

200. TiNx coatings prepared by d.c. reactive magnetron sputtering.
Musil, J., Bardos, L., Rajsky, A., Vyskocil, J., Dolezahl, B., Dadourek, K., Kubicek, V., Thin Solid Films, 136 (1986)

201. TiNx compounds deposited at low temperature: General aspects ot their phase composition,
Martev, I.N., Grigorov, G.I., Petrov, I.G., Dynowska, E., Thin Solid Films, 131 (1985)

202. TiN coatings on steel,
Buhl, R., Pulker, H.K., Moll, E., Thin Solid Films, 80 (1981)

203. TiO2 / (indium tin oxide) multilayer film: a transparent
     IR reflector,
     Sawada, Y., Taga, Y., Thin Solid Films, 116 (1984)

204. Very high rate reactive sputtering of TiN, ZrN and HfN,
     Sproul, W.D., Thin Solid Films, 107 (1983)

205. A model of silicon solar cells for concentrator photovoltaic/thermal system design,
Mbewe, D.J., Card, H.C., Card, D.C., Solar Energy Vol. 35 No. 3, (1985)

206. Analysis of spectrally selective liquid absorption filters for hybrid solar energy conversion,
Chendo, M.A.C., Osborn, D.E., Swenson, R., SPIE Vol. 562, (1985)

207. Analytical evaluation of a solar thermophotovoltaic (TPV) converter,
Edenburn, M.W., Solar Energy Vol. 24, (1980)

Abstract:
This parametric analysis of a thermophotovoltaic (TPV) converter considers emitter temperature cell reflectance to radiation with energy below the cell's bandgap energy, and concentration ratio requirements. The concentration ratio is rigorously considered to determine its influence on converter performance. Important conclusions are that an emitter temperature near 2000 K is optimal; a cell reflectance value of 0,98 is required for below-bandgap irradiation; a secondary concentrator must be used with a parabolic-dish primary; and a mirror quality resulting in a 4-mrad reflected-beam dispersion is required for a 24 per cent conversion efficiency.

208. Component optimization for photovoltaic/thermal systems,
Schwinkendorf, W.E., Solar Engineering Vol. 5, (1983)

209. Concentrating collectors for thermophotovoltaic converters, Bracewell, R.N., Price, K.M., Swanson, R.M., Solar concentrating collectors, (1977)

210. Design consideration for flat-plate-photovoltaic/thermal collectors,
Cox III, C.H., Raghuraman, P., Solar Energy Vol. 35 No.3, (1985)

Abstract:

Several potentially useful features in the design of photovoltaic/thermal (PV/T) collectors are explored in order to determine their effectiveness and interaction. Based on a computer simulation of flat-plate PV/T collectors that is applicable to a wide range of designs, the present work focuses on air-type collectors employing single crystal silicon PV cells. Features explored center on two main areas; increasing the solar absorptance and reducing the infrared emittance. The results of the simulations can be summarized as follows; for PV cells covering greater than approximately 65 % of the total collector area, a selective absorber actually reduces the thermal efficiency when used with a gridded-back cell. The requirements for the low emissivity coating are an infrared emissivity of less than 0,25 and a solar trans- missivity of greater than 0,85. The optimum combination for an air PV/T was found to consist of gridded-back PV cells, a nonselective secondary absorber, and a high- transmissivity/low-emissivity cover above the PV cells.

211. Development of the Sandia 200 X experimental silicon module, Arvizu, D.E., Conf. Rec. IEEE Photovoltaic Spec. Conf., (1984)

212. Efficient hybrid photovoltaic-photothermal solar con- version system with cogeneration,
Soule, D.E., Rechel, E.F., Smith, D.W., Willis, F.A., SPIE 562, (1985)

Abstract:

A new type of concentrating photovoltaic-photothermal solar conversion system with output cogeneration is pre- sented. This technique, called total solar cogeneration (TSC), converts the total solar spectrum directly and cogenerates the output into three energy components: high-temperature heat (HT), photovoltaic electricity (PV), and low-temperature heat (LT). A specially designed heat-mirror with a beam-splitting technique is used to direct a selected portion of the solar spectrum to a HT evacuated-tube receiver. This high-grade heat transfer is optimized, while effectively maintaining the integrity of the photovoltaic conversion efficiency. High-current A.S.E.C. silicon solar cells (nominally 18,3% efficient at 28 Grad C and 35x) are used. An analysis of heat- mirror transmittance profiles has led to an optimized theoretical parametric model profile, that in a TSC system is capable of delivering HT heat with 39% effi- ciency, while reducing PV efficiency by only 3,4%. In an experimental TSC concentrating module using a dielectric- Au-dielectric multilayer heat-mirror with optical losses, the projected output is HT heat (150 - 250 degrees C) at 17,8% efficiency, 12 V d.c. PV electricity at 9,5%, and LT heat (50 - 70 degrees C) at 41,9%, with a total cogenerated output efficiency of 69%.

213. Erstellung und Test von Experimental-Solargeneratoren für
     konzentrierende Systeme,
     Gochermann, H., Mühle, H., BMFT-Forschungsbericht
     T81-193, (1981)

Abstract:

Starting from the results of the BMFT sponsored fundamental research on silicon solar cells for concentrated
sunlight terrestrial photovoltaic experimental generators
have been designed, developed and built, which are suitable for integration into a "hybrid system". The hybrid
system comprising MAN's concentrating collectors HELIOMAN
and AEG-TELEFUNKEN's photovoltaic generators, renders
possible system tests of interacting components, which
are known from separate operation.

214. High temperature solar energy conversion system,
     Price, M.K., Solar Energy 25 (1980)

215. High temperature solar collector of optimal concen-
     tration: non-focussing lens with secondary concentrator,
     Collares-Pereira, M., O'Gallagher, J., Rabl, A., Winston,
     R., Egger, J., Williams, K., Sun: mankind's future source
     of energy Vol. 2, (1978)

216. Hybridgenerator zur Gewinnung von Strom und Wärme,
     Wegmann, H.D., Berichtsband zum 3. Internationalen VDM-
     Symposium Solartechnik, (1980)

217. Image collapsing concentrators,
     Sletten, C.J., DOE/CS/34163-2, (1980)
     Novel materials and devices for sunlight concentrating
     systems,
     Hovel, H.J., IBM J. Res. Dev., (1978)

218. Passive cooled 1000 X GaAs module with secondary optics,
     Kaminar, N., Borden, P., Photovoltaic specialists con-
     ference NY, U.S.A., (1982)

219. Performance of a combined photovoltaic/thermal flat-
     plate, liquid collector,
     Aiello, W.A. Raghuraman, P., Solar Engineering Vol. 3,
     (1980)

Abstract:

A combined photovoltaic/thermal (PV/T), flat-plate,
liquid collector, where the liquid circulates both below

and above the photovoltaic cells (the primary energy-absorbing surface), has been designed and tested according to ASHRAE 93-77 specifications to yield collector thermal and electrical efficiencies. A onedimensional thermal analysis predicts the test results accurately. On the strength of the test and analytical results, design recommendations are made to maximize the total energy extracted from the collectors.

220. Photovoltaic conversion of concentrated solar radiation,
Krebs, K., Gianoli, E., Luxemburg Commission of the European Communities, (1978)

221. Photovoltaic/thermal system sizing,
Schwinkendorf, W.E., Solar Engineering Vol. 6, (1984)

222. Photovoltaischer Hybridkollektor,
Karl, H., Bericht der Universität - Institut d. Elektrotechnik, Stuttgart, (1980)

223. Race for future electrical power- 2. Solar cells,
Tebo, A.R., Electro-Opt. Syst. Des. Vol. 12 No.11, (1980)

224. Solar cells with concentrators,
Mlavsky, A.I., Winston, R., US-Patent 4045246, (1977)

225. Solar energy collector,
Rex, D., Erno Raumfahrttechnik, Patent, (1979)

226. Spectral selectivity applied to hybrid concentration systems,
Hamdy, M.A., Luttmann, F., Osborn, D.E., SPIE Vol.562, (1985)

227. Sputtered In_2O_3: Sn-films: preparation and optical properties,
Jiang, S.J., Granqvist, C.G., SPIE Vol.562, (1985)

Abstract:

We prepared In_2O_3: Sn coatings by reactive dc magnetron sputtering. At a substrate temperature of 300-400 degrees C and a deposition rate of ~0,8 nm/s we could obtain transparent films with ~4% normal luminous absorptance, ~88% normal thermal reflectance, and ~3 x $10^{-4}\Omega cm$ electrical dc resistivity. Spectrophotometric measurements were used to evaluate the complex dynamic resistivity. It could be reconciled with the Gerlach-Grosse theory for a gas of free electrons damped by ionized impurity scattering.

228. The design and performance of ideal solar concentrators based on the prism-assisted cylindrical reflector, Edmonds, I.R., Cowling, I.R., Meara, L.A., Wheeler, B., Solar Energy Vol. 30 No. 6, (1983)

229. Thin film multilayer filter designs for hybrid solar energy conversion systems, DeSandre, L., Sond, D.Y., Macleod, H.A., Jacobson, M.R., SPIE Vol. 562, (1985)

230. Transparent heat mirror characteristics of cadmium indate, Haacke, G., SPIE Vol. 562, (1985)

MC

ENERGIE- UND UMWELTTECHNIK GMBH

CONSIDERATIONS AND PROPOSALS FOR
FUTURE RESEARCH AND DEVELOPMENT
OF HIGH TEMPERATURE SOLAR PROCESSES

F. BOESE
P.E. HUBER
H.W. KAPPLER
J. LAMMERS

MOTOR COLUMBUS, STUTTGART

Contents   Page

1. Introduction

The "Deutsche Forschungs- und Versuchsanstalt für Luft-
und Raumfahrt" (DFVLR) is in the process to define new
projects and experiments within the framework of high
temperature solar energy research. The DFVLR charged the
"MC Energie- und Umwelttechnik" to develop ideas and
proposals in this field. Special weight should be layed
upon systems with very high temperatures and upon processes with some fundamental advantages compared to systems considered so far.

2. Description of the basic assumptions and boundary conditions

In order to make intelligible, which general idea is followed by the subsequent considerations and proposals, we describe briefly the overall goals assumed for the development.

We presuppose that R+D-work based upon our considerations or proposals shall give evidence on the technical feasibility of the processes envisaged. These processes shall not be investigated to such an extent, that commercial plants can be designed with the results gained. Therefore, also new ideas with a long term development potential shall be included.

On the other hand we presume, that only technological development shall be considered, i.e. development of technical systems and components based on known physical and chemical phenomenons. Basic research on physical or chemical problems is assumed to play a supplementary role within certain parallel development tasks.

We assume within this study, that new solar systems concepts shall be primarily directed towards the lowest possible costs of the generated energy. Other aspects like "small units", "simple or appropriate technology", "fabrication chances in developing countries" are not considered here.

The requirement "lowest possible costs" does not definitely determine the optimum figure for the efficiency to be attempted. Theoretically low costs can be realized also with low efficiencies if low operational and, in particular, low investment costs are obtained as well.

The practical experience with low efficiency systems
(e.g. solar ponds or the Manzanares plant) shows, however, that a certain cost level cannot be undercut: Even
very cheap foil reflectors, foil covers or other large
area components have to be designed for hail storms and
other environmental impacts. Therefore, there seems to
be only a relatively small potential for future cost reductions with these systems. Due to this reason, we confine our considerations on high efficient systems. But
one has to state, that there is no proof in principle on
the cost disadvantage of low efficient systems.

The considerations of the subsequent sections and the
proposals made can be devided into two different groups.
The first group confines ideas and proposals, which are
specificly concerned with very high temperatures. The
reasons for these considerations are developed in chapter 3. Within the second group proposals are made, which
are not exclusively related to very high temperatures,
but may also be of interest, if very high temperature
systems will not prove feasible or economical.

Within the chapters "development programs" we estimate
also costs. It has to be stressed, that these figures
are very rough and that they are dependent on several
boundary conditions, which are not known today. For instance, the costs of computer program development depend
very strongly on the availability of codes similar to
the new problems to be solved. Or, for material testing
programs, the laboratory equipment, already existing, is
important. In all cost estimates we assume the existence
of such basic know-how or equipment, the respective work
being carried out by the DFVLR oder other institutes or
companies having available the necessary qualified sources.

3.   Basic considerations

There is quite a number of high temparature solar plants
erected and operated in the world, solar tower plants as
well as farm concepts. A broad variety of primary cooling media has been applied: water, gas, sodium, salt.
Being different in their thermodynamic concepts as well
as in many design detail, practically all of them had
one common property as far as the receiver principle is
concerned: The receivers were designed as "conventional"
heat exchanger, i.e. the radiation is absorbed at the
surface of a vessel (pipes) containing a cooling medium.
The heat generated on this surface is then transferred
via heat conduction through the wall of the vessel and
finally transferred to the cooling medium. This principle is meant by "conventional" heat exchanger. It was developed for the transfer of sensible heat from one medium to another one. Due to the fact, that solar energy
is initially not contained in a medium, but is pure
radiation, the conventional heat exchanger design is not
necessarily the adequate approach for converting sunlight into heat.

If other conversion concepts can be realized, also the
technical restrictions for conventional heat exchangers
could be overcome, e.g. with respect to maximum temperatures. Already 30 years ago proposals were made for
other concepts and some of them have been tested in
Odeillo. Recently interesting investigation including
tests was untertaken by A.D. Hunt. In the next chapter
we will describe the basic aspects and properties of
these concepts. Without going into details, which are
not part of this study, we conclude that there is reasonable hope for new receiver types, allowing much
higher temperatures than realized today with receivers
of the heat exchanger type.

Whether these new receiver types will be feasible, or
- more precisely - which temperature level will be feasible is open at present. Therefore we have to discuss
first, which temperature level is required from the
thermodynamic point of view.

3.1   Idealized processes without losses

We start with a very simplified consideration on the
basis of idealized processes. This is only to make clear
the most important aspects. Later on we will go into
more details under nonideal, realistic conditions.

Consider a surface with the area a and temperature T,
which absorbs solar radiation with the initial intensity
ϕ being concentrated by a factor c at the absorbing surface. The energy flow balance is given by

$$\dot{Q} = \alpha_s \, a \, c \phi - \varepsilon_{th} \, (T) a \sigma T^4,$$ (1)

whereas α_s is the absorption coefficient, ε_{th} (T) the
emission coefficient for the thermal spectrum represen-
ted by the temperature T, and σ the Boltzmann constant.

$\dot{Q}$ becomes larger with decreasing temperature. On the
other hand, if we assume, that the heat being generated
is used in a subsequent ideal process with constant temperature input and Carnot efficiency, the temperature of
the heat should be as high as possible. The power, which
can be generated then is given by

$$N = \dot{Q} \; \frac{T - T_a}{T}$$ (2)

T_a being the ambient temperature.

The maximum value for the power N is reached, if the solar radiation is concentrated as high as possible, i.e. c = 46000. In this case one can write

$$c\phi = \sigma T_\odot^4 \, ,$$ (3)

whereas $T_\odot$ is the temperature of the sun surface. For the power one finds then

$$N_{c\,max} = a\sigma\left(\alpha T_\odot^4 - \varepsilon T^4\right)\frac{T-T_a}{T} \, .$$ (4)

For further simplification we assume $\alpha = 1 = \varepsilon$.
(this is not necessary, because ε may be smaller than α if $T \neq T_\odot$!). The optimum temperature in order to reach the highest possible value for N_{cmax} is determined by

$$\frac{\partial N_{c\,max}}{\partial T} = 0 \, .$$ (5)

This yields an equation for T

$$4T^5 - 3T^4 T_a - T_\odot^4 T_a = 0 \, .$$ (6)

For $T_\odot$ = 6000 K and T_a = 300 K the optimum temperature is found to be T = 2545 K.

This result means: An ideal solar thermal system, optimized for maximum power output, should operate at an absorber temperature of 2545 K (or even more: due to the fact that T   T   in this case,   can be smaller than   from the physical point of view, reducing the emission term in (4) and, therefore, allowing T to be higher).

For T = 2545 K it is to be found

$$\dot{Q}_{cmax} = 0,968 \ \alpha \sigma T_{\odot}^{4} \tag{7}$$

and

$$N_{cmax} = 0,854 \ \alpha \sigma T_{\odot}^{4}, \tag{8}$$

i.e.: under idealized conditions (exception: $\mathcal{E} = \alpha$) a maximum efficiency of 85 % referring to the total radiation, not scattered by the atmosphere, can be reached.

The question is now, whether or to which extent the value for the optimum temperature will deviate from the ideal figure, if realistic systems are considered.

At first we pay regard to losses keeping for the moment the other assumptions (c = c_{max}, $\alpha = 1 = \mathcal{E}$, homogeneous temperature distribution over the absorber area, heat input into the thermodynamic process with constant temperature).

Idealized processes with losses

There are two types of losses possible in the process, heat losses and irreversible mechanisms. Heat losses can be payed regard to in supplementing equation (1)

$$\dot{Q} = \alpha_s \, a \sigma T_\odot^4 - \varepsilon_{th}(T) \, a \sigma T^4 - a \dot{q}_L \, . \qquad (9)$$

Irreversible mechanisms can be described by reducing the temperature T of the thermodynamic cycle to an effective temperature T_p. This can be represented by introducing the factor

$$P = \left(\frac{T_p - T_a}{T_p} \right) \left(\frac{T}{T - T_a} \right) \qquad (10)$$

into equation (2). With $\alpha = 1 = \varepsilon$ one can write

$$N_L = a \left(\sigma (T_\odot^4 - T^4) - \dot{q}_L \right) \frac{T - T_a}{T} \, P \, . \qquad (11)$$

T_p shall be smaller than T, i.e. $P < 1$.
We can write

$$T_p = \mu T \quad \text{with} \quad \mu < 1 \, . \qquad (12)$$

Using again the condition $\dfrac{\partial N}{\partial T} = 0$, we find now for the optimum temperature

$$\left(4\mu T - 3 T_a \right) T^4 = \left(T_\odot^4 - \frac{\dot{q}_L}{\sigma} \right) T_a \, . \qquad (13)$$

As long as

$$\mu > \frac{3T_a}{4T}$$ (14)

and $$\dot{q}_L < \sigma T_o^4 \, ,$$

which is fulfilled for all reasonable technical processes, one finds:

$$T_{opt} \left(\dot{q}_L \neq 0 \right) < T_{opt} \left(\dot{q}_L = 0 \right)$$ (15)

and

$$T_{opt} \left(\mu \neq 1 \right) > T_{opt} \left(\mu = 1 \right) .$$ (16)

This means: If heat losses occur in the system but no exergy losses due to irreversible processes, the optimum absorber temperature should be smaller than in the ideal case without heat losses. On the other hand, if only irreversible processes occur and no heat losses, than the optimum temperature is higher than in the ideal case.

As a consequence one can conclude, that only a detailed analysis of a real technical system, taking into account all types of losses, can lead to a proven judgement on the optimum absorber temperature. This is because different types of losses have contradictory influences on the value of this temperature.

3.3   First approach to real systems

As a next step from idealized to realistic technical
conditions, we pay regard to the fact, that the heat
generated at the absorber has to be removed by a heat
transfer medium. If the heat can be transferred to this
medium by exciting chemical or latent energy levels, the
heat transfer can be performed on a practical constant
temperature level. However, technically proven systems
are available only with heat transfer media using their
sensible heat. In this case the medium enters the absorber with the temperature T_1 and leaves it heated up to
the temperature T_2. As well as on the side of heat generation, a temperature difference between T_2 and T_1
is necessary on the side of heat utilization: The heat
is transferred to the subsequent thermodynamic process
between these two temperature levels.

It is well known that the efficiency of an ideal process
(without irreversibilities) can be written as

$$\eta = \frac{\overline{T} - T_a}{\overline{T}} \qquad (17)$$

the thermodynamic average temperature being

$$\overline{T} = \frac{T_2 - T_1}{\ln \frac{T_2}{T_1}} \ . \qquad (18)$$

In general the temperature distribution across the absorbing area a is not constant. Therefore equation (1)
for the energy flow balance has to be modified.

- 182 -

If da is an infenitesimal area element with temperature T one can write

$$d\dot{Q} = \alpha_s \, da \, c\phi - \varepsilon_{th}(T) \, T^4 da.$$ (19)

In order to solve this equation, one needs a relation between da and T. This is given by

$$d\dot{Q} = c_p \, \dot{m} \, dT,$$ (20)

where c_p is the specific heat and $\dot{m}$ the mass flow of the heat transfer medium at the location of the area element da.

Usually it is assumed that $\dot{m}$ is constant for all area elements. But this is not necessary! On the contrary the distribution of $\dot{m}$ over the absorption area is one of the important parameters to be optimized for a solar thermal system with high efficiency. There are several possibilities to realize non-uniform mass flow distributions technically, for two dimensional as well as for three dimensional (volumetric) absorbers.

If the mass flow distribution is determined, da is determined as function of T according to

$$da = \frac{c_p \, \dot{m} \, dT}{\alpha_s c\phi - \varepsilon\sigma T^4}.$$ (21)

- 183 -

Using (21) the heat output of the absorber with radiated area a is given by

$$\dot{Q} = \alpha_s \, a \, c \phi - \sigma \int_{T_1}^{T_2} \frac{\varepsilon \, T^4 c_p \, \dot{m} \, dT}{\alpha_s c \phi - \varepsilon \sigma T^4} \, .$$

(22)

The power output of the whole system is then given by

$$N = \dot{Q} \, \frac{\overline{T} - T_a}{\overline{T}} \, .$$

(23)

One can see, that even in this simplified consideration a clear cut judgement on the optimum temperature of the absorber is difficult. In order to develop a certain "feeling" it is necessary to make further reasonable assumptions.

For simplicity we assume a linear temperature increase across the absorption area:

$$da = \varsigma \, dT.$$

(24)

This is valid if

$$\dot{m} = \frac{\varsigma}{c_p} \left(\alpha_s c \phi - \varepsilon \sigma T^4 \right)$$

(25)

or

$$\dot{m} = \frac{\varsigma}{c_p} \left(\alpha_s c \phi - \varepsilon \sigma \left(\frac{a}{\varsigma} + T_1 \right)^4 \right) \, .$$

(26)

Inserting (25) into (22) we find

$$\dot{Q} = \alpha_s \, a \, c \, \phi - \varepsilon \sigma \varsigma \int_{T_1}^{T_2} T^4 \, dT \qquad (27)$$

or $\qquad \dot{Q} = \alpha_s \, a \, c \, \phi - \varepsilon \sigma \varsigma \frac{1}{5} \left(T_2^5 - T_1^5 \right) , \qquad (28)$

whereas ς is determined by integrating (24):

$$\varsigma = \frac{a}{T_2 - T_1} \, . \qquad (29)$$

With (29) equation (28) reads

$$\dot{Q} = \alpha_s \, a \, c \, \phi - \varepsilon \sigma a \, \frac{1}{5} \, \frac{T_2^5 - T_1^5}{T_2 - T_1} \qquad (30)$$

or $\qquad \dot{Q} = \alpha_s \, a \, c \, \phi - \varepsilon \sigma \frac{a}{5} \left(T_2^4 + T_2^3 T_1 + T_2^2 T_1^2 + T_2^3 T_1 + T_1^4 \right). \quad (31)$

In order to obtain an evaluation of this formula, we investigate the meaning of the term $\frac{1}{5}(...) = \tilde{T}^4$ for an example. Let us assume T_2 to be about 50 % higher than T_1:

$$T_2 = \frac{3}{2} \, T_1 \, . \qquad (32)$$

- 185 -

Then we find

$$\tilde{T}^{4} = \frac{1}{5}(\ldots) = \frac{1}{5}\left(\frac{81}{16} + \frac{27}{8} + \frac{9}{4} + \frac{3}{2} + 1\right)T_{1}^{4} = 2{,}637\ T_{1}^{4}. \quad (33)$$

On the other hand, for the thermodynamic mean temperature $\bar{T}$ the following figure results with the assumption (32):

$$\bar{T} = 1{,}233\ T_{1}$$

or $\qquad \bar{T}^{4} = 2{,}312\ T_{1}^{4}.$ (34)

If we replace $\tilde{T}^{4}$ in equation (31) by $\bar{T}^{4}$, the heat output value is reduced by about 12 %. This reduction is of the order of magnitude of heat losses $\dot{Q}_{L}$ (e.g. by convection and conduction) from the high temperature system components of real systems. Therefore, we can write for a system with heat losses:

$$N = \left(\dot{Q}(\tilde{T}) - \dot{Q}_{L}\right)\frac{\bar{T} - T_{a}}{\bar{T}} \approx \dot{Q}(\bar{T})\ \frac{\bar{T} - T_{a}}{\bar{T}}. \quad (35)$$

This formula compares exactly with equation (2), and the consideration with respect to the optimum temperature is exactly the same as in (5) and (6), T being replaced by $\bar{T}$. Instead of (6) the optimization results in the equation

$$4\bar{T}^{5} - 3\bar{T}^{4}T_{a} - \frac{\alpha c \phi}{\varepsilon \sigma}\ T_{a} = 0. \quad (36)$$

It can be seen, that the value of the optimum mean temperature T_{opt} depends on the ratio $\dfrac{\alpha \, c \, \phi}{\varepsilon}$.

Before we start the discussion on the influence of this factor if non-ideal figures for c and ϕ are taken into account, we go back on the assumption (24) or (25), respectively. Because the equations (28) to (36) are only valid, if this assumption can be made, we have to prove it.

The physical meaning of the equations (24), (25) and (26) is, that the mass flow $\dot{m}$ is not homogeneously distributed over the absorbing area. For a two dimensional as well as for a three dimensional (volumetric) absorber this can be technically realized. For a two dimensional absorbing area very arbitrary distributions can be reached by varying the thickness of the channels containing the heat transport medium. As long as a good heat transfer, e.g. by a turbulent flow, can be provided, any geometry can be choosen.

For volumetric absorbers it is more difficult to provide arbitrary flow geometries. But also in this case non-uniform distributions can be realized.

As a final remark to this problem we would like to stress, that the distribution choosen with (24) to (26) is not necessary. It was only made in order to simplify the consideration. When designing systems, this flow distribution has to be optimized. One can imagine that even more favourable distributions than given by (26) can be provided for. Then smaller radiation losses are to be expected, and the value for the optimum absorber temperature (similar to T) will become higher.

Volumetric receivers could be designed in that way that
the region with the highest temperature is far from the
window. In this case the high temperature re-radiaton is
most likely to be trapped.

Now we continue with the problem of the optimum mean absorber temperature T_{opt}, but now with non-ideal
figures for $c\phi$.

3.4   Second approach to real systems

The following relations between $\overline{T}_{opt}$ and c can be calculated using equation (36):

$\alpha/\varepsilon = 1$

$\emptyset$ \ T_{opt}	1000	1200	1400	1600	1800	2000	2200	2400
1300	451	1173	2624	5237	9618	16516	26910	41968
1100	533	1387	3101	6190	11366	19519	31802	-
900	651	1695	3790	7565	13892	23856	38870	-
700	837	2180	4873	9726	17861	30672	-	-
500	1172	3051	6822	13617	25006	42941	-	-
300	1953	5085	11370	22695	41676	-	-	-

$\alpha/\varepsilon = 2$

$\emptyset$ \ T_{opt}	1000	1200	1400	1600	1800	2000	2200	2400
1300	225	587	1312	2619	4809	8258	13455	20984
1100	266	694	1550	3095	5683	9759	15901	24799
900	325	848	1895	3783	6946	11928	19435	30310
700	418	1090	2436	4864	8931	15336	24988	38970
500	585	1526	3410	6809	12503	21470	34983	-
300	975	2543	5684	11349	20838	35784	-	-

$\alpha/\varepsilon = 5$

$\emptyset$ \ T_{opt}	1000	1200	1400	1600	1800	2000	2200	2400
1300	90	235	525	1047	1923	3303	5382	8394
1100	106	278	620	1238	2273	3904	6361	9920
900	130	340	758	1513	2778	4771	7774	12124
700	167	437	975	1945	3572	6134	9995	15588
500	234	612	1364	2723	5000	8588	13993	21823
300	390	1020	2274	4539	8334	14313	23322	36372

In order to evaluate these figures, we have to discuss the chances of realizing favourable figures for $\frac{\alpha}{\varepsilon}$ and c, and also, which figure has to be assumed for ϕ.

Let us start with the last question. Outside the atmosphere in vicinity of the earth one finds $\phi = 1300$ W/m^2 whereas at the surface of the earth, there is even under favourable conditions a maximum intensity of direct solar radiation of about 1000 W/m^2 and the daily average being smaller. It seems reasonable to investigate the problem for space and terrestrical application separately:

a) space applications

If one considers solar thermal power stations for extra terrestrical application, one can assume, that very large parabolic concentrators can be realized with optimum and permanent orientation to the sun. Even with simple mirrors made of light weight plastic material, concentration factors of 5000 to 10000 can be obtained. For thoroughly fabricated rotational symmetric parabolic concentrators values up to 20000 have already been demonstrated. This figure still represents less than 50 % of the theoretical limit.

If we assume 50 % of the theoretical value (c = 46000) as technical/economical limit for space applications a mean temperature of about 2200 K should be realized for the input even for $\alpha/\varepsilon = 1$, i.e. $T_1 = 1784$ K and $T_2 = 2676$ K. Due to the fact that the spectrum of the solar radiation and the emission spectrum do not overlay very much even in this temperature region, there is the theoretical possibility that a certain selectivity can be realized. In this case even higher temperatures are required for optimum efficiency, or this temperatures are optimum even with smaller concentrations.

It is interesting, that already a relatively small selectivity of $\frac{\alpha}{\varepsilon} = 2$ leads to an optimum mean temperature of $T_{opt} = 1800$ K for a concentration factor of less than 5000. This might not be a very hard technical requirement. In this case one should design the system for $T_1 = 1460$ K and $T_2 = 2190$ K.

We would like to stress, that for space systems high inlet temperatures are decisive, because this is a prerequisit for high temperatures of the waste heat of the process to be radiated to space. If this latter temperature is not high enough, the area of the radiator becomes large, and therefore, the radiator becomes too heavy.

b) terrestrical applications

For terrestrical systems local weather conditions as well as the day/night cycle is of major importance for the value of the optimum inlet temperatures. An optimum process should be designed to operate with a changing heat inlet, i.e. absorber temperature. Technically, this might be very difficult to realize. But one can imagine that the system could be sub-divided into two sub-systems. One system could operate on a lower temperature level during the morning and evening hours, whereas the high temperature system is operated during the hours around noon (4 - 6 hours?), delivering its waste heat to the "low temperature" process.

In this case for the high temperature sub-system a
mean radiation intensity level during operation of
more than 700 W/m^2 might be assumed. Whereas the
average value for a one step system may vary around
500 W/m^2.

Today concentration factors of about 1000 are envisaged for solar tower plants. Looking to the tables
above one can see, that for this concentration factor
and a solar radiaton intensity of about 500 W/m^2 a
mean inlet temperature optimum of about 1000 K is required, if one assumes, in addition, that no selectivity can be obtained (α / ε = 1). That means for
instance T_1 = 811 K and T_2 = 1216 K, if we keep
the assumption $T_2 = \frac{3}{2} T_1$ as made in the preceed-
ing sections.

For the evaluation of future possibilities it is very
important that the above assumption is discussed. It
might well turn out after a certain phase of research
and development, that high concentration factors as
well as selectivity is technically not feasible or
economically not adjustable. But on the basis of our
present knowledge this is not proven, nor even investigated. One should realize that a concentration
factor of 1000 represents only 2 % of the theoretical
figure! Technical and commercial systems usually obtain a much better approximation to theoretical limits.

This consideration is also valid for the selectivity
of the absorber. Due to the relatively small overlaps
of the spectra of incoming and re-emitted radiation,
selectivity is physically possible. Selective coatings may not be applicable in high temperature systems. But with volumetric receivers, selectivity may

be obtained by designing the absorbing and heat transfer system in a proper way. For instance, if absorbing particles are mixed into a gas, the volumetric absorption coefficient in the spectral region of the incident solar radiation may be choosen small, i.e. this radiation is absorbed within a certain distance (absorption length) by the gas flow. On the other hand the volumetric absorption coefficient for the wavelengths of the thermal re-radiation may be choosen high (e.g. by adding a second type of absorbing particles). Than trapping of the re-radiated light is to be expected /3.1/, and selecitivity occurs.

Therefore, it might be reasonable to allow higher values for the concentration also for terrestrical applications, as well as a certain selecitivity. Consider for instance $\frac{\varkappa}{\Sigma}$ = 2 and a concentration of 12000, than even for ϕ = 500 W/m^2 an optimum mean temperature of T_{opt} = 1800 K is required, i.e. T_1 = 1460 K and T_2 = 2190 K.

As final remark we would like to point out, that a concentration ratio of about 15000 has been demonstrated in Odeillo. High concentrations are technically feasible, if necessary by twofold concentrating systems. Whether their application is reasonable has to be proven by future system optimization and economical considerations. We have also to refer to the fact, that we made numerous simplifications of the real situation in the preceding sections. The assumptions were made in order to emphazize the basic aspects. But one has to have in mind, that for instance the concentration is usually not homogeneously distributed over the absorbing area, or that the mass flow distribution can be choosen in another way as done above.

3.5   Additional considerations and conclusions

The present stage of high temperature solar power plant
development can be thought to be represented by a "GAST"-
concept. There is a gas cooled heat exchanger type of
receiver with an inlet temperature of about 600 K and an
outlet temperature of about 1100 K. Near term improvements are attempted by increasing the inlet and outlet
temperature to about 700 K/1300 K, using ceramic material. In order to become successful with this second
stage of development, severe material problems have to
be solved. In this case, the mean temperature of heat
input into the thermodynamic process would be increased
from about 825 K to about 970 K. If we assume for the
"present" as well as for the future process a concentration factor of about 1000, the overall efficiency of the
ideal reference processes is increased from about 62 %
to about 65 %, i.e. a theoretical improvement of about
5 % could be expected. The heat loss due to re-radiation
from the absorber increases from about 3 % to about 6 %.

Let us compare this theoretical potential of improvement
of the overall efficiency with the theoretical potential
of improvements of optimum high temperature system with
mean temperatures of 1600 K or 1800 K, as considered in
the preceding chapters. The respective figures are de-
picted in the following table:

	"GAST"	"GAST" improved	high temperature I	high temperature II	
T	825	970	1600	1800	K
C	1000	1000	7565	13892	-
radiation loss 1)	3	6	6	5	%
overall efficiency	62	65	77	80	%
efficiency improvement 2)	-	5	25	30	%

1) in percent of the incident radiation
2) compared to the theoretical efficiency of an ideal
   "GAST"-process (related to the power of the incident
   radiation)

The figures of this table show on the one hand, that even in case of a successful material development program with heat exchanger types of receivers no substantial improvement of the overall efficiency can be expected. If, on the other hand, volumetric absorbers with mean temperaturs of 1600 K to 1800 K and high concentrations and/or a certain selectivity can be realized, the theoretical potential for the efficiency improvement ranges from 25 to 30 %.

But this theoretical potential may be only one aspect. It might be possible (and should be investigated) that in the very high temperature region also other types of thermodynamic cycles could be realized, which show practical advantages over "medium"-temperature concepts. If, for instance, the number of sub-systems is smaller, the overall process concept is simpler, the storage and part load properties are better, it might well be that the theoretical potential could better be exhausted by real technical systems. This might be one of the most important aspects for thermochemical cycles for hydrogen production, which will be discussed in chapter 7.

3.6 Research proposal

The discussion above has shown a number of basic thermo dynamic questions. As far as we know, they have not yet been investigated to the necessary depth. We suggest thermodynamic and design studies in particular with the following questions:

- influence of losses onto the optimum temperature
- optimization of the mass flow distribution over the absorption area
- possiblities to realize $\frac{\alpha}{\epsilon} > 1$ with volumetric absorption
  ° technical concepts
  ° thermodynamics of trapping mechanisms
- possibilities and influence of high concentrations
  ° technical concepts
  ° cost trade offs for different concepts
  ° influence on the value of the optimum temperature, including non-uniform distributions

We estimate the reasonable effort for these investigations of about 4 man-years.

4.   Receiver concepts

With respect to the possible temperature range, which
can be reached in the connected thermodynamic cycles,
two different types of receivers can be realized:

Indirect heat generation

The solar light is not directly absorbed by the cooling
medium, but it is absorbed at the outer surface of pipes
containing this medium. The generated heat is transferred via heat conduction through the pipe walls into the
cooling medium.

This is the construction principle, which was applied in
practically all test and pilot plants during the past
ten years. With this concept fundamental material problems are inevitably connected, even with ceramic material. With the present technology and available material
temperatures up to 1300 K (of the cooling medium) seem
to be possible, long term resistance to the solar specific permanent transients not proven by practical experience. There is a very small development potential for
very high tempratures with this concept. Even on the
long run 1500 K may be a target very hard to obtain.

Direct heat generation

The solar light is directly absorbed by the cooling medium (or by absorbing material, which is contained in
the cooling medium, and which transfers the generated
heat instantaneously to the cooling medium). The light
is preferably absorbed volumetric. The cooling medium
may (but must not necessarily) be separated from the ambient air by a transparent cover.

The respective receiver types have been investigated up
to now only to a relatively small extent, e.g. about
30 years ago by Odeillo researchers, and recently by research groups in Perpignan, Toulose, SERI, Berkeley and
Seattle. There is no practical experience of prototypes
connected with thermodynamic cycles. Theoretical investigations as well as some experimental results /4.1 to
4.4/ show, that most likely there will be much less temperature restrictions due to fundamental material problems:

- with open systems temperatures up to 3000 K can be
  realized
- window temperatures (with quartz) of about 1100 K seem
  to be feasible today (if severe transients can be
  avoided)
- outlet temperatures (of systems with windows) of more
  than 2000 K seem to be no serious problem.

There is also the theoretical possibility of realizing a
certain selectivity ($\frac{\alpha}{\varepsilon} > 1$) with volumetric absorption.

The principles and problems of receivers shall not be
discussed in this study. We restrict ourselves on the
statement that direct absorber type receivers have much
more favorable properties with respect to very high temperatures, than heat exchanger type receivers. We recommend therefore to investigate direct absorption (or volumetric absorption) receivers in greater detail. The
investigation should include liquid and gaseous as well
as solid absorption media.

For the consideration in the subsequent chapters we
assume, that volumetric absorbers will be investigated

in a research program, in order to obtain very high temperatures up to more than 2000 K, at least for the receiver outlet temperature.

The considerations of chapter 2 have shown, that if receivers can generate such high temperatures one should also try to realize thermodynamic cycles on this temperature level. Whether there are chances for the technical realization of very high temperature processes is discussed in chapter 5 and 6. Respective proposals are developed.

5.   MHD-systems with gas as thermodynamic and electric
       working medium

5.1   Fundamental aspects

We assume the function principles of MHD-generators to
be known, and we restrict ourselves on the discussion of
those aspects and properties, which are relevant with
respect to the questions to be considered here.

The physical feasibility of MHD-systems with gas as
working medium has been proven more than 20 years ago.
The reasons why they are not yet applied technically or
commercially are different for the two different construction principles. Usually one distinguishes between
open and closed loop processes. But the fundamental differences between them are not mainly due to the property
"open loop" or "closed loop". Rather, in open loop processes flue gas is assumed as working medium, whereas in
closed loop processes, one assumes noble gases to be
applied. It is the choice of the working medium that is
decisive for the feasibility, efficiency and costs of a
MHD-system: flue gas shows much less electrical conductivity than noble gases. Or more specific: noble gases
show the same conductivity figures for much less temperatures (see fig. 5.1). Noble gases, therefore, are much
more favourable to be used in MHD-processes.

On the other hand a certain minimum value for the conductivity has to be provided for (about 0,1 S/cm) at the
inlet of the MHD-channel. For noble gases this figure is
reached not below a temperature of more than 2000 K, even
if seed material (cesium) is added (see fig. 5.1 and 5.2).

conductivity S/cm

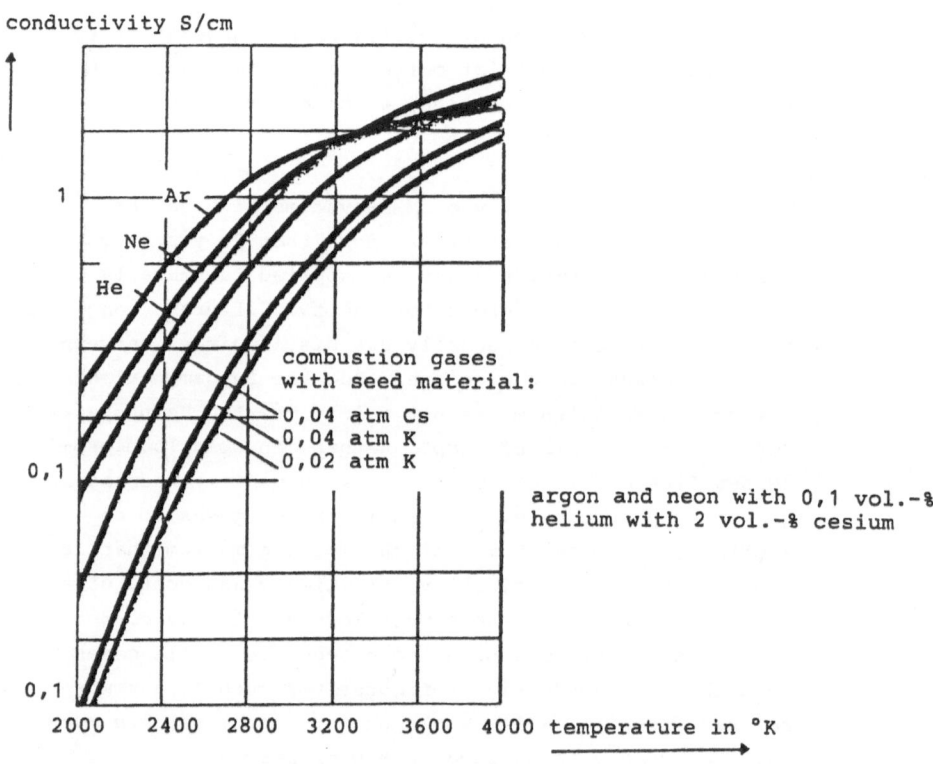

Fig. 5.1: direct current conductivity of gases with
seed material, pressure 1 kp/cm^2

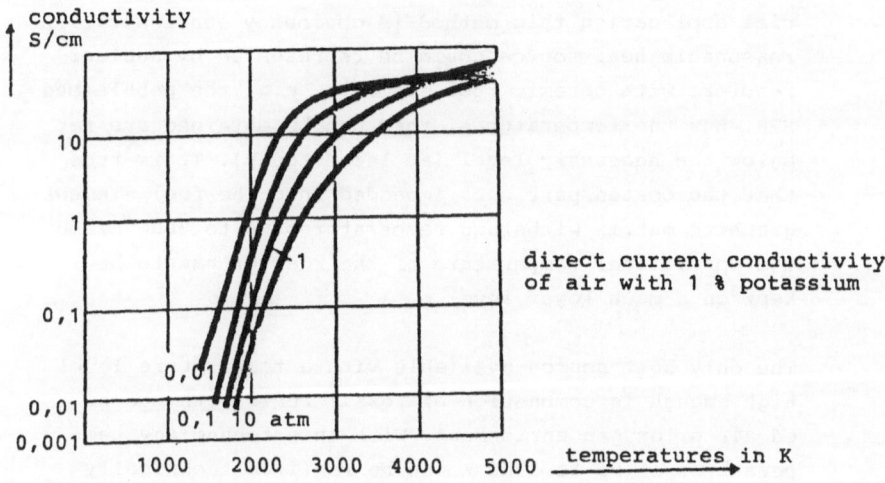

Fig. 5.2: direct current conductivity of air with
1 % potassium at different pressures

And here the reason for the unsolved problems of MHD-
systems shows up: to date there is no reasonable energy
source with sufficient temperature available, which can
be used for the heating of a noble gas circuit.

Heating can be provided for with electrical heaters (resistance heating). With this method experimental circuits have been realized and operated. But for commercial application this method is obviously nonsensical. A
reasonable heat source could be represented by nuclear
reactors with ceramic fuel elements, e.g. the pebble bed
HTR. But the temperatures, that can be obtained are far
below the necessary level (at least today). It is true
that the coated particles imbedded into the fuel element
graphite matrix withstand temperatures up to 2000 K, but
the operational temperature of the reactor has to be
kept on a much lower level.

The only heat source available with a temperature level
high enough is combustion of fossil fuels with pre-heat-
ed air or oxygen enrichment. With this technology temperatures of up to 3000 K can be realized technically
today. In order to use this high temperature heat of a
combustion chamber within a MHD noble-gas-cycle the heat
had to be transferred from the flue gas to the noble
gas. This transfer is technically feasible today only on
a temperature far below 3000 K, perhaps up to 1300 –
1400 K.

It was necessary to describe these facts to a certain
extent, because they were causal for the development of
MHD-systems in the past and even up to now. The development of noble gas systems was terminated after a period

of basic investigations. The basic research had shown
that the MHD-generator itself would be feasible, but no
possibility was seen for a reasonable heat source.
Therefore, the research was concentrated on the development of MHD-systems with flue gas as working medium,
which have necessarily to be open loop systems. All
scientists were aware that in this case temperatures of
up to 3000 K had to be controlled.

The development has shown that one can realize systems,
which can be operated under technical conditions, and
which may reach sufficient long lifetimes. But on the
other hand it turned also out, that one needs very com-
plicated devices and very expensive material to ensure
the technical feasibility. For instance, there is always
a segmental cooling system necessary, and in the inlet
area of the channel platin as material is used. With
present fuel prices, there is no justification for MHD-
systems as topping processes in power plants. The increase of efficiency does by far not compensate the additional investment costs.

The situation can be summarized as follows:

° MHD-channels with flue gases need inlet temperatures
  of about 3000 K
  - heat source (combustion of fossil fuel) available
  - but severe material problems with the MHD-channel
    itself

° MHD-channels with noble gases need "only" inlet temperatures of about 2100 K
  - no fundamental material problems with the MHD-chan-
    nel itself
  - but up to now no heat source available (high tempe-
    rature nuclear reactor supplies "merely" about 1200 K).

5.2   Concepts for solar application

If we consider the aspects described in the chapters 3,
4 and 5.1 it is obvious to think about the combination
of a volumetric solar receiver (with noble gas as
cooling medium) with a closed loop MHD-cycle. Recent
theoretical and experimental investigations with
different direct absorption receivers have shown, that
gas temperatures of more than 2000 K might be feasible.
This fits quite well with the requirement for the inlet
temperature of a noble gas MHD-channel.

Therefore, a system as shown in the flow diagram in fig.
5.3 seems to be interesting. As bottoming process a combined gas turbine/steam turbine cycle could be considered with system parameters similar to the thermodynamic
cycle of the GAST-concept.

The thermodynamic principle of the whole process is
shown in fig. 5.4 Approximate main parameters are to be
found in table 5.1.

If for the receiver window silica glass is used (e.g.
Corning Vycor Code No. 7913), it might be possible to
realize window temperatures in the vicinity of the
"cold" gas inlet of about 1100 K. If saphir is used a
window temperature of about 1300 seems to be the maximum
value. The "cold" gas flowing away from the window area
into the inner region of the absorber could be heated by
solar radiation via suspended absorbing particles up to
more than 2100 K, if the hot gas flow does not contact
the window.

The MHD-channel outlet temperature should reach about
1300 - 1400 K.

H   heat source   MHD   heat transformer   HT   heat transfer
P   pipe   D   diffuser   C   compressor

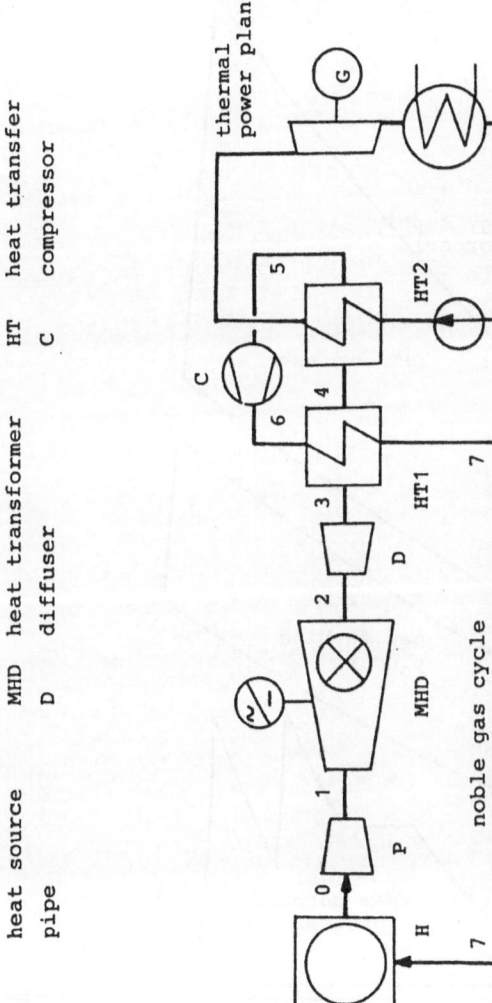

Fig. 5.3: basic flow diagram of a closed MHD-plasma-process

- 207 -

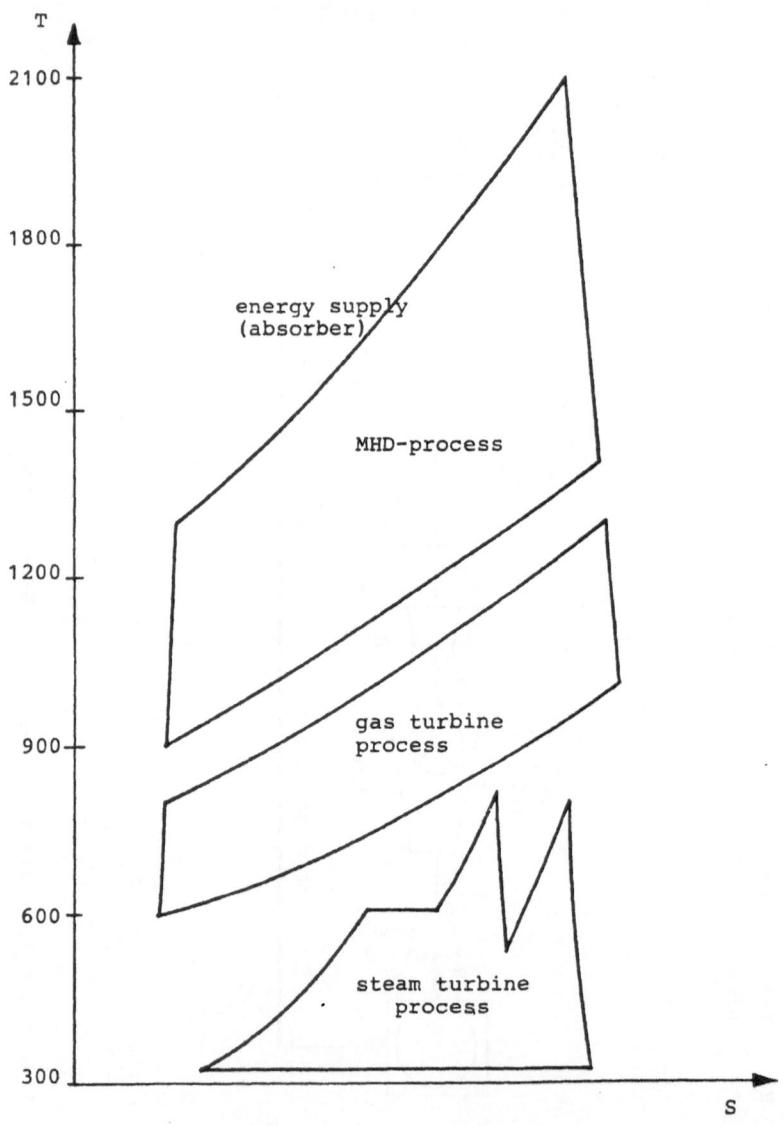

Fig. 5.4: T, s-diagram of a (3-step) solar MHD-
gas and steam boiler process (no optimization carried out)

		temperature
	maximum temperature receiver	2100
	MHD-inlet	1400
I	heat transfer-outlet	900
MHD	compressor-outlet	1300
	heat transfer-outlet	1300
II	turbine-outlet	1000
gas-	waste-heat-boiler gas outlet	600
cycle	compressor-outlet	800
III	live steam temperature	900
steam-	condensator temperature	300
cycle		

Tab. 5.1: approximate main data of a closed MHD-
cycle with downstream gas and steam cycle

The further process flow, behind the MHD-channel, depends on the pressure level, which can be realized with
the MHD-process, i.e. which pressure the receiver can
withstand. If the initial pressure can be made high
enough, the waste gas from the MHD-channel could directly brouhgt into a gas turbine, avoiding an intermediate
heat exchanger (a prerequisit being that neither the
particles within the gas necessary for the light absorption nor the particles necessary for the conducitivity
are damaging the turbine). Behind the turbine the de-
pressurized gas will then deliver its residual heat to a
subsequent steam turbine process. Finally, the gas is
compressed again to the receiver pressure level. Systems
optimization might lead to equal power ratings for the
gas turbine and the compressor, i.e. the power of the
gas turbine is completely transferred to the compressor,
there is no generator connected to the turbine. The
total net output power is solely generated by the MHD-
generator. A schematic flow diagram of this process is
shown in fig. 5.5.

If there is no sufficient pressure in the receiver, and
therefore also behind the MHD-channel, the heat of the
MHD-waste gas can only be transferred by a heat exchanger to a separate gas turbine cycle, which is on a
higher initial pressure level.

For the MHD-generator itself the following main parameters
are likely to be favourable:

- argon as working medium
- cesium as seed material
- thermal power density (in the channel itself) about
  10 MW/m^3

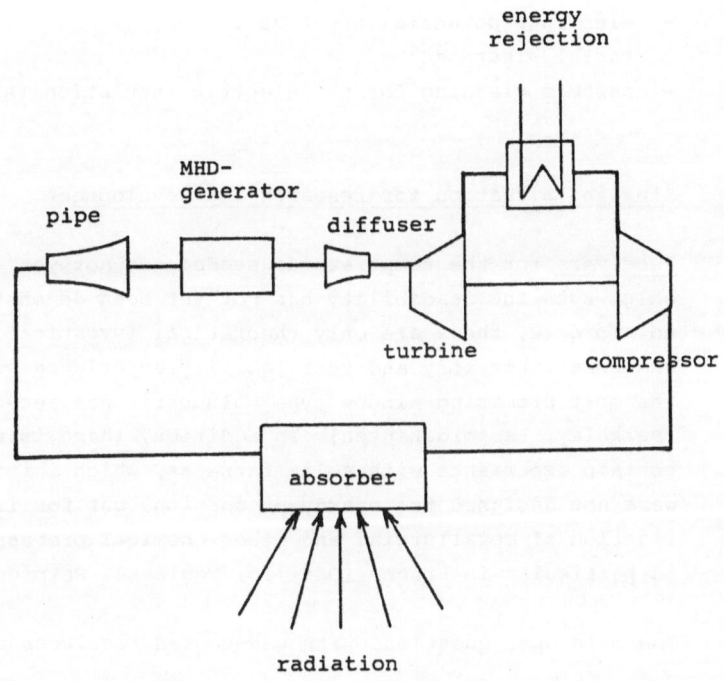

Fig. 5.5: basic flow diagram of a MHD-gas turbine process
with net electricity output solely from MHD-generator

- flow velocities
  ° head channel about 30 m/s
  ° main nozzle about 700 m/s
- inlet pressure 10 bar
- electrode potential about 20 V
- tantal electrodes
- ceramic cladding for the electric insulation (Al_2O_3).

5.3   Starting situation for research and development

Receivers for the temperatures needed are not yet available. Even the feasibility has not yet been demonstrated. To date, there are only theoretical investigations
and some laboratory and test facility experience with
the most promising window type volumetric gas receiver
(Berkeley, Lampoldshausen). In addition, there is also a
certain experience with solar furnaces, which initially
were not designed for energy production, but for investigation of metallurgial and other chemical processes,
in particular in France (Odeillo, Toulouse, Perpignan).

The main open questions with gas-cooled receivers are:

- Which maximum temperature at which pressure is obtainable ?
- Which minimum (inlet) temperature is necessary with
  resepct to window life time ?
- Is seed material (necessary for the electric conductivity) compatible with the window material ?

A further interesting question might be, whether solar
radiation could increase the conductivity at the MHD-
channel inlet. If, for instance, cesium is used as seed
material, the ionization spectrum and the solar spectrum

show a considerable overlap, and thus, solar radiation will increase the conductivity as far as the light can penetrate far enough into the channel region, where this conductivity is required. In principle, this seems possible, if the channel is directed versus the radiation beam, and is arranged immediately behind the volumetric receiver. The ratio of ionized to neutral particles could reach 10^{-6} to 10^{-5} depending on the pressure, the temperature and the concentration factor of the light.

If the conductivity within the channel could be increased, the channel inlet temperature could be reduced.

MHD-system with noble gases were investigated also in Germany. Research was terminated in the early seventies due to the lack of a reasonable heat source. Experts opinion was that the channel itself would be technically feasible. There are still some major components available, which were used within the research program in Jülich. In particular, a test channel for 10 kW power output exists in Essen (Prof. Bohn), which could be brought into operation with relatively small effort. Present value of the channel: about 1 million DM.

Also still available is one of the leading experts in the field, Prof. Bohn. If MHD-research would be continued, he is interested to co-operate within a respective program.

A broad research on MHD-systems has been executed in the Argonne National Laboratory. The program is continued up to now. Major interest was given to open cycle systems. But in the past years also several "non-conventional" concepts have been investigated.

The main questions with a solar powered noble gas MHD-channel are:

- Which minimum inlet temperature is necessary ?
- Which maximum outlet temperature is necessary ?
- Can cooling of the channel be avoided ?
- Is the absorption material (necessary for solar radiation absorption within the receiver) compatible with the channel walls as well as the seed material (short circuits ect.) ?

For the system of receiver and channel, the following questions can be seen immediately:

- optimum inlet and outlet temperatures, evaluating the advantages/disadvantages of the influences on the receiver and the channel
- compatibility of seed and absorption medium within the gas.

5.4   Description of development goals

Based upon the situation as described in chapter 5.3 the following near term development targets can be derived:

- MHD-generator

  ° Verification of former experimental results and investigation of possible improvements, taking into account the progress in material development of the past 10 years.

  ° Special development of concepts for low inlet/outlet temperatures

° development of systems, which are not critical with
  respect to transients (which are typical for solar
  systems, and which have not yet been studied in the
  necessary depth)

- Receiver

  Development of volumetric receivers with the following properties:

  ° Argon as cooling medium
  ° outlet temperature as high as possible (up to 2200 K)
  ° Inlet temperature as high as possible (more than
    1200 K)
  ° high pressure (about 10 bar)
  ° seed material within the cooling gas (preferably
    cesium)

- System

  ° optimized receiver and MHD inlet/outlet tempera-
    tures and pressures
  ° amplification of the conductivity within the MHD-
    channel by residual solar radiation
  ° good part load properties
  ° optimum adaption of the MHD-cycle to the following
    thermodynamic process.

5.5   Description of critical problems

Most likely the most critical problem is the resistance
of the receiver window against

° high temperatures
° temperature shocks
° high pressure
° sedimentation, diffusion and corrosion due to seed and
absorption material

Critical for the MHD-channel is whether or not wall
cooling will be necessary.

5.6   Main elements of a possible development program

Due to the fact, that one can not predict whether the
necessary temperatures can be provided with volumetric
receivers, research has to concentrate first on this
question. On the other hand, whether one should try to
reach this goal at all, is depending on the feasibility
of a respective MHD-system. Therefore, also parallel research on this question should be carried out.

We suggest that receiver research should be intensified
on the theoretical as well as on the experimental side.
Test facilities on a size of about 100 kW_{th} should be
erected and operated under solar conditions, e.g. in
Almeria.

MHD-research should concentrate in a first step on theoretical investigations in order to collect and to evaluate all available data on the worldwide research over
the past ten years and to find out the bandwidth of possible parameters. A respective study might need about
2 - 4 man years.

Experimental research should be started only after a
careful evaluation of the study results and after encouraging results from the receiver tests. MHD-test facilities are expensive, even if some components of former
research could be used. As an example: An electric heater for the test channel available at the institute of
Prof. Bohn would cost about 3 million DM.

We estimate the volume of a reasonable starting program
to a minimum of 10 million DM.

5.7 First evaluation of development chances

The feasibility of a solar MHD noble gas cycle will
mainly depend on the feasibility of the high temperature
receivers. For this major component no evaluation of the
chances is possible today. On the one hand, there are
some fundamental advantages of volumetric receivers with
windows over heat exchanger type receivers, but on the
other hand temperatures of more than 1200 K combined
with thermal transients might be a very hard boundary
condition for the window, even with cooling devices.

If the necessary temperatures can be provided by a respective receiver, all other problems, including those of
the MHD-generator, seem to be less severe.

6. MHD-systems with gas as thermodynamic and liquid metal as electric working medium

6.1 Fundamental aspects

Even ionized gas of 2000 to 3000 K is much less conductive than liquid metal. There is a factor of about 10^4 between the respective conductivities. Therefore, also liquid metal MHD-systems were investigated. The result of the early research was, that a liquid metal MHD-generator could easily be realized. But as a major problem there was always a very pure efficiency of the transformation of heat into kinetic energy of the liquid metal. All processes suggested up to the late seventies were highly irreversible.

In one kind of processes jet propulsion played a major role (see fig. 6.1 and 6.2). This is already by basic physical principles a very irreversible process, even if the injection is realized in several steps.

Another group of processes is represented by two phase flows for acceleration of the liquid metal, and separation of the flows before the MHD-channel in order to have a pure (one phase) liquid flow within the channel. There are no basic physical principles requiring irreversibility. But it turned out that it is technically impossible to avoid high losses of kinetic energy for fast flow.

It was only in the late seventies, when the idea was investigated in more detail, to allow two phase flow still within the MHD-channel, and to separate the flows behind the channel, when their velocity is low (Branover, Pierson et al). In this case much lower thermodynamic losses (i.e. transformation losses from heat into kinetic

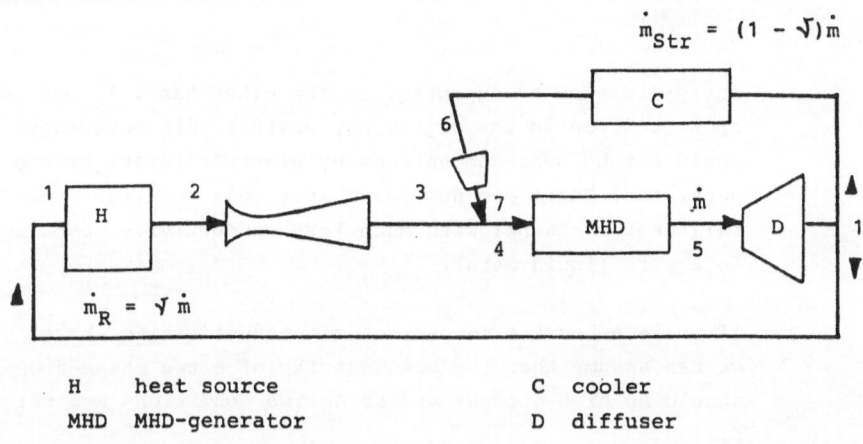

$$\dot{m}_{Str} = (1 - \sqrt{})\dot{m}$$

$$\dot{m}_R = \sqrt{}\dot{m}$$

| H | heat source | C | cooler |
| MHD | MHD-generator | D | diffuser |

Fig. 6.1: basic flow diagram of a MHD-process with jet
propulsion

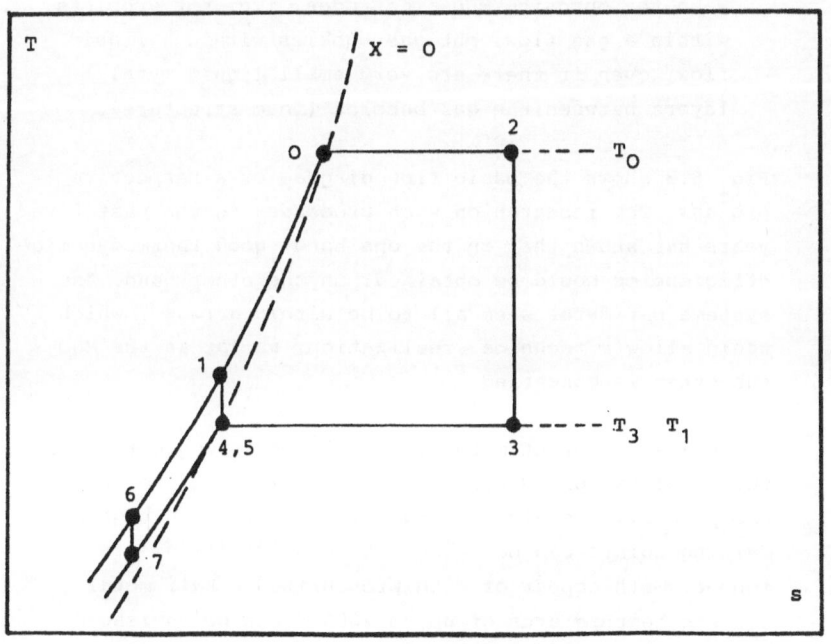

Fig. 6.2: T, s-diagram of a MHD-process with jet propulsion

energy) are to be expected. On the other hand, it was an
open question in the beginning, whether this advantage
would not be over-compensated by electric losses of the
generator: There are now two phases (gas and liquid metal) in the channel with much less conductivity compared
to a pure liquid metal.

If we remember the comparison of conductivities above,
we can assume that the conductivity of a two phase flow
should be high enough, as far as two conditions are fulfilled:

- overall density of the two phase medium not smaller
  than 10^{-4} times the density of pure liquid metal

- non-interrupted connection of liquid metal films between the opposite MHD-electrodes, i.e. not droplets
  within a gas flow, but gas bubbles within a liquid
  flow, even if there are very small liquid metal
  layers between the gas bubbles (foam structure).

Fig. 6.3 shows the basic flow diagram of a respective
process. The research on such processes in the past five
years has shown that on the one hand, good thermodynamic
efficiencies could be obtained. On the other hand, the
systems parameter seem all to be within a range, which
would allow a technical realization, as far as the MHD-
sub-sytem is concerned.

There are different metals, which come into question for
this application. Usually alkali metals (e.g. sodium)
are primarily considered. But also metals with higher
melting points can be taken into consideration, e.g.
copper. With copper or with pressurized alkali metal
systems temperatures of up to 2000 K can be envisaged.
Therefore, one could think of a combination of these
kind of systems with high temperature solar receivers.

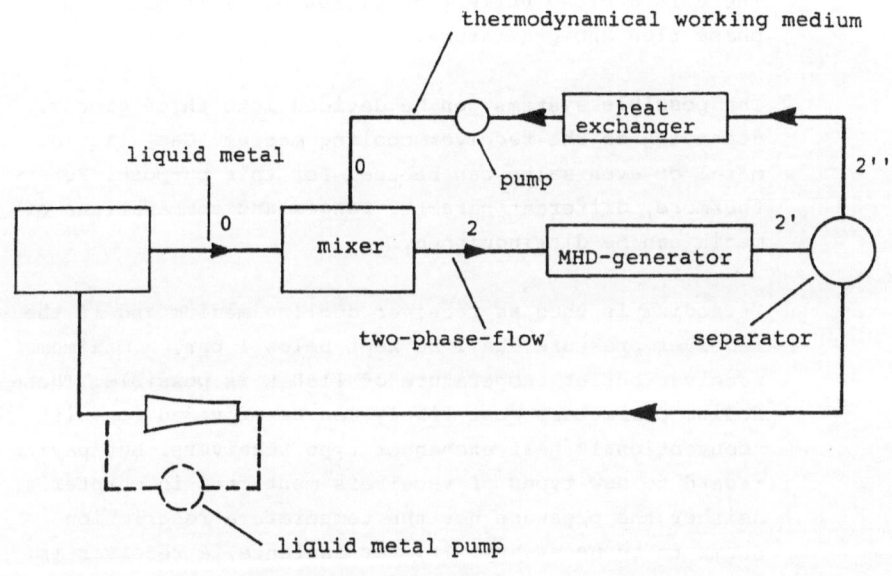

Fig. 6.3: basic flow diagram of a MHD-process with
emulsified two-phase-flow and separation
behind the MHD-channel

6.2   Concepts for solar application

There is a broad variety of cylces possible with two
phase flow MHD-generators.

The possible systems can be devided into three groups,
depending on the receiver cooling medium. Gas, liquid
metal or even salts can be used for this purpose. Furthermore, different paramter ranges and combinations of
media can be distinguished.

If sodium is used as receiver cooling medium and if the
receiver pressure shall be kept below 1 bar, a maximum
receiver outlet temperature of 1150 K is possible. These
sodium parameters most likely can be provided for with
"conventional" heat exchanger type receivers. But paying
regard to new types of receivers mentioned in chapter 4,
neither the pressure nor the temperature restriction
seems to be necessary. If, for instance, a receiver is
considered, where a sodium flow is realized behind a
window, but without direct contact of window and sodium,
a pressure of about 10 bar and an outlet temperature of
1500 K might be possible. In this case the solar radiation is absorbed directly by the sodium.

With the same receiver concept but with copper as cooling medium (theoretically) possible outlet temperatures
range from 1360 K to 2500 K if the receiver pressure
shall be kept below 1 bar.

It is obvious that also mixtures of liquid metals can be
used as cooling media. In this case, both, pressurized
systems without evaporation as well as systems with
steam generation can be considered.

One can see that already the variety of liquid metal receiver types opens a broad field of basic questions. It is not possible within the framework of this study, to discuss merely one of these possibilities in greater detail. We have to confine ourselves on a pure itemizing of the different aspects and concepts. This is also valid for the system variants, being related to the different receivers.

6.2.1 Systems with liquid metal receivers

Let us consider now one system variant, showing the main aspects of liquid/gas systems, namely a sodium receiver with a pressure of 10 bar and an outlet temperature of say 1300 K. If the receiver outlet flow with this parameters is mixed with, for instance, a noble gas of 1300 K and 10 bar within the mixer, one can expand than this two phase mixture into a MHD-generator. Due to the fact that the sodium has a much higher specific heat capacity than the gas, and that the gas mass compared to the sodium mass can be kept small, a nearly isothermal expansion can be realized. Therefore, the thermodynamic cycle transforming heat into work is quite different from a usual gas turbine cycle.

The expansion within the channel can be performed until a pressure of about 3 bar is reached. Lower figures would lead to evaporation of sodium. In this case, one part of the sodium could be separated only after condensation of this sodium fraction. If a final pressure of 3 bar is realized, the two media can be separated completely behind the MHD-channel. The sodium flow (with only small temperature decrease compared to the receiver outlet temperature) is then pumped back to the receiver, e.g. by an electro-magnetic pump. The gas flow is cooled

by a heat exchanger to about 1000 K, and then compressed
again to 10 bar and 1300 K.

The application of this sodium MHD-cycle as topping process to gas turbine cycle leads to a thermodynamic improvement, though the maximum temperature (1300 K) is
not increased: The heat input into the sodium receiver
occurs between a relatively small temperature range Δ T,
i.e. the Carnot factor is no longer determined by the
mean inlet temperature of the gas turbine cycle but by
the nearly constant temperature between inlet and outlet
of the receiver (or the MHD-channel respectively). The
disadvantage of the system is, that a relatively high
sodium mass flow has to be circulated in order to absorb
and transfer the necessary heat power.

In order to increase the maximum temperature, on the one
hand, sodium systems with higher pressure have to be
realized, or one has to choose metals with lower evaporation pressures, e.g. copper. On the other hand, the
thermodynamic cycle of the working gas has also to be
changed: Due to the fact, that efficient recuperative
heat transfer from the working "MHD-gas" to a bottoming
process is technically not feasible for temperatures
above about 1300 K, the MHD-gas process needs an adiabatic expansion step behind the isothermal expansion.
This can be realized by injecting "cold" liquid metal
into the MHD-channel and thus reducing the main stream
liquid metal temperature in an adequate way. The "cold"
liquid metal has to be provided for by a second liquid
metal cycle with a heat exchanger on a temperature level
below 1300 K. For this purpose the liquid metal flow behind the separator is divided into a main flow being directed to the receiver, and an auxiliary flow being directed to the cooler or to the injector, respectively.

For the inlet temperature of the working gas into the
mixer there are two possibilities. One solution is to
compress the gas behind the heat exchanger to the initial pressure and to a temperature of about 1300 K. A
higher temperature is technically not feasible this way.
The gas is then heated up by the liquid metal, using the
high heat capacity of this liquid metal. After this
heating procedure, an isothermal expansion takes place,
and subsequently an adiabatic expansion.

The respective temperature and pressure development can
be controlled within the MHD-channel by a proper design
of the magnetic forces, the channel geometry and the injection of "cold" liquid metal.

The other possibility for designing the hot gas end of
the gas cycle is to provide a second receiver for the
gas, besides of the receiver for the liquid metal. In
this case also the gas could be heated up to the same
level as the liquid metal, e.g. to about 2000 K with a
volumetric gas receiver.

6.2.2   Systems with gas receivers

The last example of the preceding section was already a
system with a gas receiver. However, the main flow of
this system was heated by a liquid metal receiver.

But there is also the possibility to introduce the heat
into the system solely by a gas receiver. In this case,
the main flow is represented by the working medium cycle. The liquid metal acts only as an auxiliary medium.
In the mixer the gas is introduced into the liquid metal
on a temperature level higher than that of the liquid
metal. The latter is first heated up (isobaric) by the

gas. Subsequently an expansion is provided for by a respectively designed MHD-channel. During this expansion, the heat supplied first from the gas to the liquid metal is now transferred back to the gas, allowing a nearly isothermal expansion.

In a next step the working gas - after separation from the liquid metal - is cooled down to less than 1300 K by a heat exchanger, delivering the waste heat to a bottoming process. After that, the gas can again be compressed to the receiver inlet status, i.e. about 10 bar and 1300 K.

6.3 Starting situation for research and development

Receivers

Liquid metal (sodium) receivers of heat exchanger type have been investigated up to a receiver outlet temperature of about 800 K. Direct absorption type receivers have not been investigated. There was also no basic research on pressurized systems (in Almeria only the necessary pressure to overcome the system friction losses were applied).

Receivers with copper as cooling medium were not investigated. But there is technical knowledge on closed liquid copper loops, also on electro-magnetic copper pumps (e.g. at INTERATOM).

MHD-systems

There is braod theoretical and experimental knowledge on liquid metal MHD-systems, mainly with sodium as electric working medium but also with sodium/potassium, mercury or lead. Also in Germany extensive research was carried out about 15 years ago (Radebold, Rex and others).

Specific know-how on two phase MHD-systems is available
at Argonne (Petrick, Piersson and others) and in Ber-
Sheva (Branover).

Copper MHD-systems have been investigated theoretically
at Argonne. They were designed as open cycle systems
with fuel gas from coal combustion as working gas within
a two phase MHD-flow.

The development of liquid metal MHD-channels has shown,
that most likely channels (or generators) would most
likely be technically feasible up to a temperature level
of about 2000 K. The main residual question is, whether
a two phase flow, i.e. the transformation of heat into
kinetic energy, will be possible with high efficiency.

6.4   Description of development goals

The discussion of the chapters 6.1 and 6.2 has shown
that there is a broad variety of possible system designs. Therefore, the main development goal should be to
evaluate the different basic cycles and to meet a decision for one or two systems to be investigated in
depth.

Besides of this decision, further work is necessary in
any case on the thermodynamic drive for the MHD-fluid.
In this respect the main goals are:

- low slip of bubbles against liquid

- high generator efficiency with high void fraction

- high temperature level of the heat input into the
  thermodynamic process

- small exergy losses due to heat transfer within the MHD-process and due to heat transfer to the bottoming process

- small separation losses liquid/gas behind the MHD-channel.

6.5   Main elements of a possible development program

The development should concentrate on three major fields:

a) Development of non-conventional liquid metal receivers (no heat exchanger types) with the main problem areas
   - compatibility of the liquid metal with the confining vessel material, in particular interaction with the window
   - direct radiation absorption
   - high pressure

b) Development of high efficient thermodynamic drive systems for the MHD-fluid
   - slip mechanisms
   - heat transfer gas/liquid
   - expansion strategies

c) Overall systems investigation
   - thermodynamic simulations and evaluations

We estimate the research effort necessary for a respective program as follows:

a) thermodynamic analysis
   4 scientists for 5 years   200 man-months

b) laboratory scale experiments
   6 scientists for 5 years   300 man-months
   budget for experiments   - 2 Mio DM -

c) testing program under solar conditions
   6 engineers and scientists for 3 years   180 man-months
   test facilities and operation costs   - 8 Mio DM -

6.6   First evaluation of development chances

For systems with sodium as working medium and temperatures up to about 1200 K there are very good chances for
technical feasibility as well as for good efficiencies.

The development of pressurized systems with sodium is
more demanding. The main open questions concern new receiver types. The MHD-channel itself seems to be less
problematic.

An evaluation on copper systems is not possible today.
Many of the related material problems on the receiver
side as well as on the side of the MHD-channel are open.
The investigation of copper systems has to start with
these basic material questions, and has to be continued
only after encouraging results on these questions as
well as on positive results from sodium systems, concerning the thermodynamic drive mechanism.

7. Thermochemical cycles for hydrogen production

7.1 Fundamental aspects

General

Storage, transport and conversion capabilities are
essential for the generated energy if the generation
plant and the consumers are locally apart. Electricity
has advantages in respect to conversion and subsequent
application but hydrogen is preferable concerning
storage and transport.

Subsequently we consider the thermochemical production
of hydrogen, where heat is used for the decomposition of
favourable chemical raw material.

Direct decomposition of water does not occur significantly below a very high temperature level of more than
5000 K. As shown in chapter 3 this temperature level is
not optimum in solar thermal systems with respect to a
maxiumum total efficiency, even if the process could be
realized technically. But obviously, the process cannot
be carried out technically on this very high temperature
level.

One possibility to overcome the problems was discussed
by Funk and Reinstrom already 1966 /7.1/. They suggested
to split the decomposition of water into at least two
reaction steps:

$$XO + heat = X + 1/2\ O_2$$
$$X + H_2O = XO + H_2$$

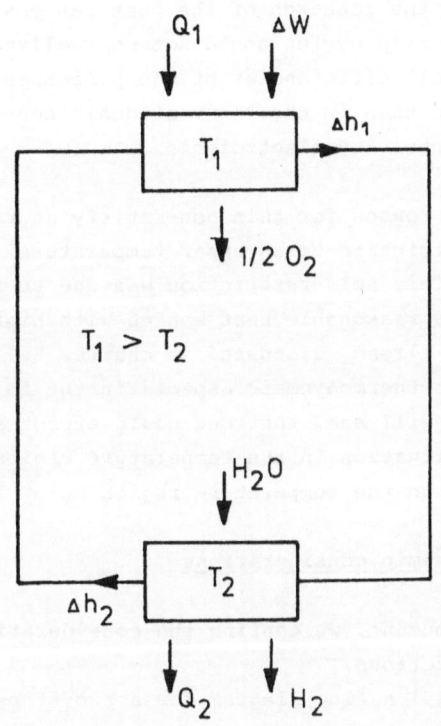

Fig. 7.1: Basic flow diagram for a two step process

However, the research of the past ten years has shown,
that two step cycles could not be realized, and that
the overall efficiencies of the processes investigated
were less than in the "conventional" conversion chain
heat-electricity-electrolysis.

The main reason for this non-satisfying result was the
self-restriction to an upper temperature level of about
1300 K. This self-restriction was due to the assumed
lack of a reasonable heat source with higher temperatures as already discussed in chapter 5.1. By discussing
the basic thermodynmaic aspects in the following sections we will see, that one could expect a complete different situation in the temperature region around 2000 K
compared to the temperature region below 1300 K.

Thermodynamic considerations

For the moment, we confine the consideration on reversible reactions.
In fig. 7.1 a flow diagram for a two-steps process is
shown. The chemical reactions shall be isobaric and isothermal on the temperature levels T_1 and T_2 where T_1
is higher than T_2. For the decomposition of water in
hydrogen and oxygen Δi and Δs are the enthalpy and entropy changes and Δi_i and Δs_i the respective ones for the
sub-reactions. At temperature T_1 we allow technical
work Δw to be added to the system.
The application of the first and the second law of thermodynamics leads then to the following equations:

$$Q_1 + Q_2 + \Delta w = Q \qquad (1)$$

$$\Delta S_1 + \Delta S_2 = \Delta S \qquad (2)$$

$$\Delta S_1 = \frac{Q_1}{T_1} \qquad (3)$$

$$\Delta S_2 = \frac{Q_2}{T_2} \qquad (4)$$

$$\Delta i = T_p \Delta S = Q \qquad (5)$$

T_p is the pyrolysis temperature of water (5400 K for steam).

By separating the decomposition of water into several steps we assume the necessary temperatures T to be lower than T_p, i.e.:

$$T_i < T_p \qquad (6)$$

Introducing T_p into (1) to (5) it is to be found:

$$Q_1 = \Delta i \, \frac{T_1 \left(T_p - T_2 \right)}{T_p \left(T_1 - T_2 \right)} - \Delta w \, \frac{T_1}{T_1 - T_2} \qquad (7)$$

$$Q_2 = - \Delta i \, \frac{T_2 \left(T_p - T_1 \right)}{T_p \left(T_1 - T_2 \right)} + \Delta w \, \frac{T_2}{T_1 - T_2} \qquad (8)$$

$$\Delta S_1 = \Delta i \, \frac{T_p - T_2}{T_p \left(T_1 - T_2 \right)} - \Delta w \, \frac{1}{T_1 - T_2} \qquad (9)$$

$$\Delta S_2 = - \Delta i \, \frac{T_p - T_1}{T_p \left(T_1 - T_2 \right)} + \Delta w \, \frac{1}{T_1 - T_2} \, . \qquad (10)$$

If we confine ourselves on the case without additional technical work, we can set

$$\Delta W = 0 . \qquad (11)$$

Due to $\Delta \dot{i} > 0$ and $T_1 > T_2$ we find:

$$Q_1 > 0$$
$$Q_2 < 0 . \qquad (12)$$

The reaction at the higher temperature level T_1 is endothermic (energy supply), the reaction at the lower level T_2 is exothermic (energy rejection).

Furthermore we find:

$$\Delta S_1 > 0$$
$$\Delta S_2 < 0 . \qquad (13)$$

The entropies ΔS_1 or ΔS_2 as functions of the temperatures T_1 and T_2 are shown in fig. 7.2 and 7.3. It can be seen that the entropy differences ΔS_1 and ΔS_2 for low temperatures T_1 are much higher than the entropy increase of the direct water decomposition. On the contrary ΔS_1 and ΔS_2 increase if T_2 increases. This means: Only for sufficiently low T_2 and sufficiently high T_1 there will be a chance to find substances that are suitable for a two-step process. Investigations /7.2 to 7.10/ have shown that at least 4 steps are necessary for $T_1 < 1300$ K and for $T_2 > 300$ K to achieve the necessary entropy increase.

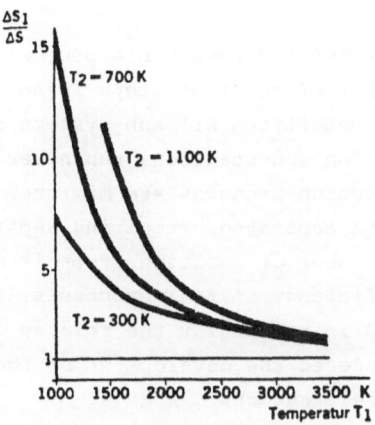

Fig. 7.2: Required entropy increase for a high temperature reaction (temperature T_1) of two step
thermochemical decomposition of water (stand-
ardized to the entropy increase of the direct
decomposition of water)

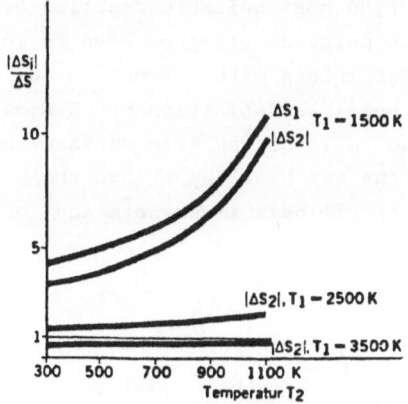

Fig. 7.3: As fig. 7.2, however, for high and low temperature reaction and dependent on the temperature T_2 of the low temperature reaction

Technical aspects

If merely idealized and reversible processes are considered, the number of reaction steps is not important.
But under real conditions all sub-systems show losses.
In particular, the sub-reactions run never complete, and
usually the reaction products are mixtures of gases,
which have to be separated, requiring separation work.
Therefore the number of reaction steps is decisive for
the overall efficiency of real processes. It was this
complexity, and in particular the related separation
loss, which hindered the development of thermochemical
cycles to become a success.

Paying regard to this situation, we can see, that the
combination of high temperature solar receivers (1300 K
to about 2000 K) with thermochemical cycles for hydrogen
production seems to be very interesting, yielding fundamental advantages.

In order to find most suitable reaction cycles from the
thermodynamic point of view, we have to look first to
favourable parameters with respect to the overall efficiency. One possible definition of this overall efficiency is the ratio of the free enthalpy of the hydrogen
produced to the net heat input into the process (i.e.
the balance of all heat flows into and out of the process):

$$\eta = \frac{\Delta g_{H_2}}{q_{in} - q_{out}} \ . \tag{13}$$

For the pre-selection of very high temperature cycles
this formula might not be reasonable, because there

might be no technical possibility for recuperation or
other utilization of waste heat on a very high temperature level. Therefore, for a first evaluation of suitable processes one should leave out of the consideration
q_{out}.

The value of q_{in} can be found for each cycle by exa-
mining the respective T-s-diagrams. An example is given
in fig. 7.4. q_{in} is given by the area below the path
1, 2, 3, 4, 5. The area below the path 5 to 10 represents the waste heat, which in principle could still be
used, but most likely has to be rejected to a certain
extent for very high temperatures. Favourable cycles are
represented by a small value for this area, i.e. dT/ds
should be as high as possible between the cycle steps
5 to 10.

7.2   Concepts for solar application

General aspects

For a "two reaction step" process having the energy
supplied at about 2000 K and removed at 400 K, the theoretical efficiency would be near 80 %.

It is obvious that such a process seems basically very
suitable to be utilized in a high-temperature solar
plant. If the problem could be solved technically, to
absorb the solar radiation at these high temperatures
into the reaction medium, high efficiencies should be
achievable.

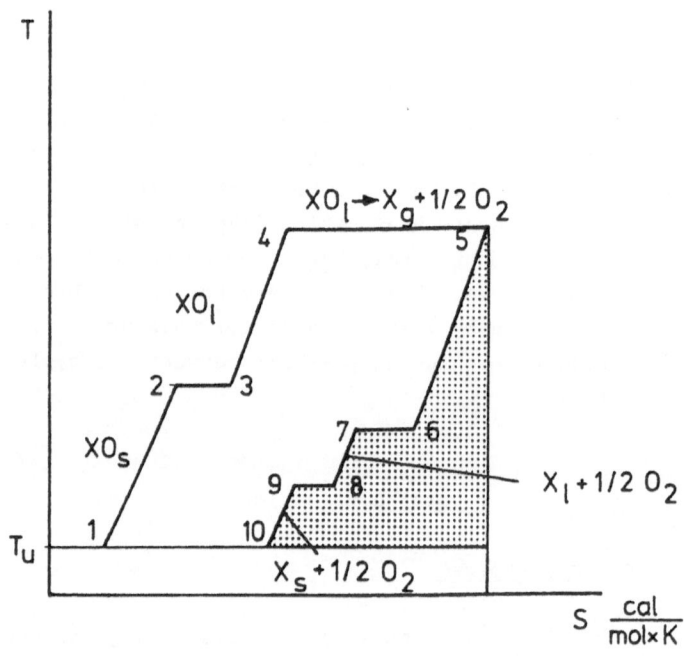

Fig. 7.4: basic thermodynamic parameters of a two step
process

Possible thermochemical cycles

An example for a "two-step"-process that seems to be
very promising is the ironoxide process

$$Fe_3O_4 \text{ (1)} \xrightarrow{\text{2000 K}} 3 \text{ FeO (1)} + 1/2 \text{ } O_2 \text{ (g)}$$

$$3FeO(s)+H_2O(g) \xleftarrow{\text{750 K}} Fe_3O_4(s) + H_2 \text{ (g)}$$

where the indices 1, s and g describe the physical state
solid, liquid and gas.

This process and other deoxidation reactions are illustrated in the temperature-entropy-diagrams and are shown
in fig. 7.5 to 7.9.

In table 7.1 these oxidation reactions are listed and
their endothermic reaction temperatures and efficiencies
are presented.

Conceptual ideas for new cycles

a) Receiver reactions

An interesting possibility for solar thermochemical
cycles might be to execute the high temperature reaction directly within a volumetric receiver, i.e.
the material to be decomposed being also the absorbing medium. The problem in connection with the absorber (receiver) is dealt with in chapter 4 as far as
necessary for conceptual considerations of new cycles.

The following functions have to be fulfilled by the
design of such a volumetric absorber

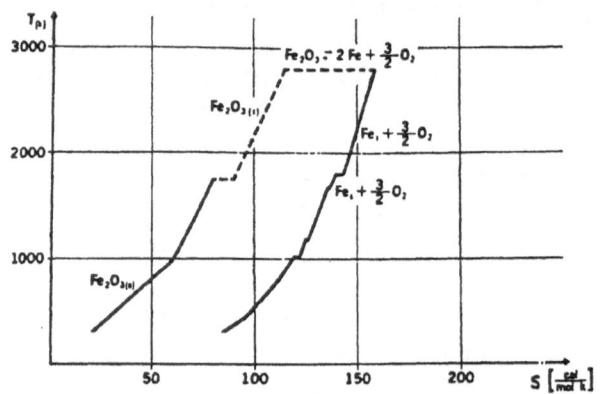

Fig. 7.5: T,s-diagram of oxide-cracking-reactions at
very high temperatures (Fe_2O_3)

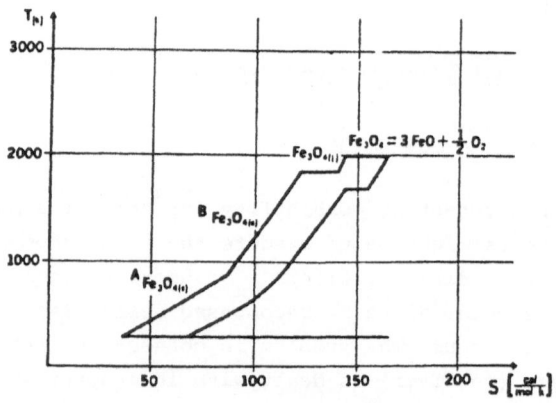

Fig. 7.6: T,s-diagram of oxide-cracking-reactions at
very high temperatures (Fe_3O_4)

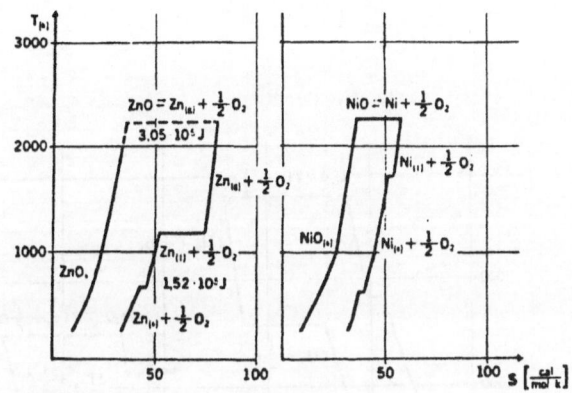

Fig. 7.7: T,s-diagram of oxide-cracking-reactions at
very high temperatures (ZnO-NiO)

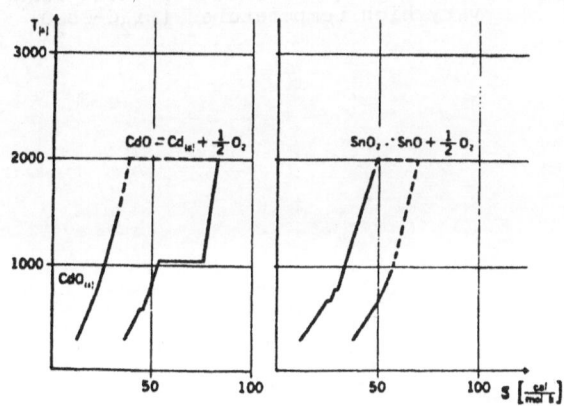

Fig. 7.8: T,s-diagram of oxide-cracking-reactions at
very high temperatures (CdO-SnO$_2$)

- 241 -

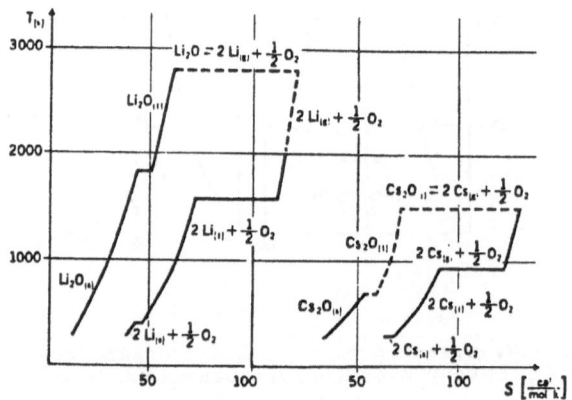

Fig. 7.9: T,s-diagram for oxide-cracking-reactions at
very high temperatures (Li_2O-Cs_2O)

reaction		temperature of the reaction *)	efficiency
Fe	$Fe_2O_3 - 2 Fe + 3/2O_2$	2.800	0,27
Fe	$Fe_3O_4 - 3 FeO + 1/2O_2$	2.000	0,45
Ni	$NiO - Ni + 1/2O_2$	2.200	0,87
Zn	$ZnO - Zn + 1/2O_2$	2.100	0,51
Cd	$CdO - Cd + 1/2O^2$	2.000	0,59
Sn	$SnO_2 - SnO + 1/2O_2$	2.000	0,9
Cs	$Cs_2O - 2 Cs + 1/2O_2$	1.500	0,65
Li	$Li_2O - 2 Li + 1/2O_2$	2.800	0,27

*) the reaction temperature can be reduced with impairment of the conversion and velocity of the reaction

Tab. 7.1: endothermic reaction temperatures and efficiencies

- containment of the material to be decomposed
- absorption of the solar radiation
- separation of the reaction products.

There are different possibilities to obtain these goals. The first is to realize open absorbers (no separation to the ambient air). The absorption medium should be liquid (or even solid).

The endothermic reaction in an open volume happens under atmospheric conditions with air being in contact with the medium to be composed. This means for the chemical reactions: The partial pressure of oxygen from the reaction must be higher/equal than the pressure of oxygen in the atmosphere. In this case the oxygen can escape into the atmosphere.

Further boundary conditions for such processes in direct contact with the atmosphere are: The absorption medium must be chemical consistent to the elements of air (CO_2, SO_2, H_2O etc.). After a certain starting phase stable conditions must be reached. Impurities could be accepted in the circuit if stable after the starting phase.

The vaporization of reacting medium into the open volume is the next criterion for losses. It is governed by the vaporization pressure of the medium. Vaporization rate, loss of materials, rate of concentration, etc. should be subjects of experiments.

Finally, the absorption of the reacting medium must be high enough, if necessary by adding absorbing particles.

The second possibility for a "receiver reaction" is
to provide a transparent cover, which has the following advantages:

- chemical compatibility between absorption medium
  and air is not necessary
- the pressure in the reaction volume can be different from the atmospheric pressure
- the absorption medium can also be a gas

The following disadvantages have to be kept in mind:

- the window absorbes and reflects a part of the radiation
- the window must be resistent to high temperatures
  and temperature transients
- the window material must be chemical resistent to
  air and absorption medium

The detailed discussion of receiver design concepts
is not part of this study. It is obvious that these
concepts are crucial, if the high temperature reaction is directly carried out within the receiver
volume.

b) Absorption and reaction in separate vessels

In this concept a receiver with gas, liquid (or even
solid material) as discussed in chapter 4 provides
heat for the endothermic reaction to be carried out
within a separate reaction vessel. This concept might
be advantageous if the boundary conditions for efficient absorption and efficient reaction are too different, and cannot be satisfied within the same
volume with the same media.

7.3   Starting situation for research and development

Since Funk and Reinstrom 1966 /7.1/ proposed and investigated theoretically the generation of hydrogen with
thermochemical cycles, many cycles that seemed suitable
have been studied in detail.

But as mentioned above, the investigation was restricted
on temperatures below 1300 K.

Due to this research on "lower" temperature processes,
there is extensive thermodynamic, chemical and process
engineering experience available, which might be very
useful and a sound basis for the investigation of high
temperature processes, in particular as far as the exothermic reactions on low temperature levels are concerned.

On the high temperature level (above 1300 K) there is no
experience on cycles. But some theoretical research was
carried out, and some specific questions of high temperature reactions under solar radiation were investigated
experimentally, in particular in France (Odeillo, Perpignan, Toulouse).

The starting situation for a R+D Program is best characterized by the following main open questions:

- Which cycles are appropriate in regard to thermodynamic considerations, i.e. pressure, temperature, efficiency, recuperative heat transfer etc ?

- Which cycles are appropriate in regard to chemical
   and technical considerations, i.e. reaction mechanism, enthalpy, physical state of reaction products,
   rate of vaporization etc ?

- Is the absorption behaviour of substances in the
  cycle favourable at high temperatures for the absorption of solar radiation ?

- Which cycles will show low separation work ?

- Is it possible to include a thermodynamical cycle for
  energy recovery at very high temperatures (heat
  transfer) ?

7.4   Description of development goals

Absorber

A receiver concept has to be developed that absorbes
solar radiation at high temperatures being able to
achieve the following:

- absorption at temperatures up to 2000 K (or even
  more)
- separation of reaction products
- maximum absorption of the radiation, avoiding reflection
- complete containment of reaction products

System

- search for low-step cylces
- development of a reaction vessel for the exothermic
  reaction
- set up of specifications, particularly chemical reactions, pressures, viscosity, absorption behaviour,
  endothermic and exothermic reactions

- adjustment of temperatures i.e. outlet temperature of
  step 1 = inlet temperatures of step 2, or if not possible: provision of a respective heat sink etc.
- system control
- part load efficiency
- simple procedures for start up and shut down operations

7.5   Description of critical problems

The critical problems are listed below:

Absorber

Open system

- problems of losses of heat and mass to the air
- entering of impurities into the process
- problem of vaporization of liquid substances
- additional chemical reactions

System with window

- heat resistent window for high temperatures
- chemical compatibility of the windows to the reaction
  products
- deposition of particles at the window
- gas separation problems

Overall system

- problems of start up and shut down operations
- stabilizing of operating conditions
- material problems of circuit and components including
  corrosion
- problems of deposition, clugging, scaling etc.

7.6   Concept of a possible development program

A possible development program can be subdivided in the
following topics:

- search of cycles and set up of basic data by theoretical and experimental investigations

- selection of cycles in regard to the main thermodynamic properties

- overall systems analysis with existing computer programs; but extension to the high temperature region
  (1300 K up to about 2000 K)

- laboratory experiments (with electric heating) for the
  crucial high temperature reactions

- design, erection and operation of complete test cycles
  on laboratory scale (with electric heating) of the
  high temperature reaction

- design, erection and operation of test facilities for
  the high temperature reaction under solar conditions
  (e.g. in Almeria)

- design, erection and operation of a complete test
  cycle under solar conditions (e.g. in Almeria).

We estimate the research effort necessary for a respective program as follows:

a) thermodynamic analysis
   3 scientists for 5 years   150 man-months

b) laboratory scale experiments
   4 scientists for 5 years   200 man-months
   budget for experiments   - 1 Mio DM -

c) testing program under solar conditions
   5 engineers and scientists for 3 years   150 man-months
   test facilities and operation costs   - 5 Mio DM -

7.7   First evaluation of the development chances

The short term research on very high temperature thermochemical cycles is related to basic chemical questions. An evaluation of the technical feasibility of complete cycles is not possible today.

Most important, also for these systems, will be the successful development of favourable high temperature receivers.

8.   High temperature latent heat storage systems (LHS)

8.1   Fundamental aspects

For energy storage in high temperature solar energy application, latent heat storage systems have been proposed and investigated. We, therefore, suppose the general properties of latent heat storage systems to be
known.

As introductory remarks for the subsequent proposals we
restrict ourselves on a summary of the problems of those
systems, which showed up so far, as well as on some fundamental advantages to be expected in the high temperature region. The main problems were:

- heat transfer problems during the period of energy
  extraction due to freezing and solidification of
  storage material around the heat exchanger tubes
- damage of the heat exchanger bundles or headers due
  to periodical volumetric change of the storage medium
- chemical instability and stratification of the storage medium and of seed crystals after a certain
  number of cycles.

In the low temperature region two further disadvantages
occur:

- relatively small value for the heat of fusion,
- relatively high ΔT-losses due to the twofold heat exchange during charging and discharging of the system
  compared to the temperature level of the stored
  heat.

Due to the fact that the value of the heat of fusion is approximately in a linear dependence on the melting temperature, for high temperature systems much larger melting enthalpies can be found. Therefore, latent energy systems, which can be designed practically pressureless also in the high temperature region, are of much more interest for application in this region, if one can solve the technical problems, which usually occurred with systems investigated in the low emperature region.

In the subsequent sections we will develop a proposal for such a system.

8.2 Description of a new design for a high temperature latent heat storage system and its basic properties

The principle design is shown in fig. 8.1 and fig. 8.2. A closed vessel contains a salt as storage medium as well as a heat transfer medium. The latter is, due to a different specific weight, separated from but in direct contact with the salt. Normally, the heat transfer medium is lighter than the storage medium, and it forms a layer upon it. Within this layer a heat exchanger is provided. Another heat exchanger is at the bottom of the vessel. Similar systems have already been proposed. One specific design possibility of the system considered here, compared to those already proposed, is that the heat can be transmitted from the surface of the molten salt to the heat exchanger pipes at the top of the vessel exclusively by natural convection. In this case, no active systems to force a flow of liquids are necessary.

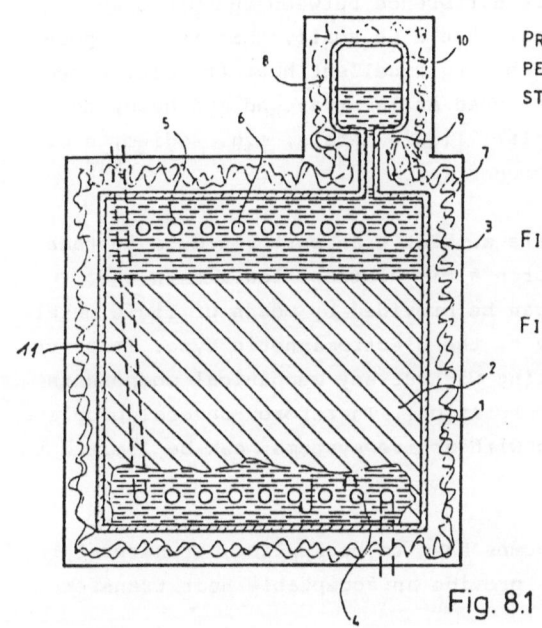

PROPOSAL FOR A HIGH TEM-
PERATURE HEAT OF FUSION
STORAGE TANK

FIG. 1. BEGINNING OF
CHARGING

FIG. 2. BEGINNING OF
DISCHARGING

Fig. 8.1

LEGEND:

1 STORAGE TANK WALL
2 STORAGE MEDIUM
  (SALT-MELT)
3 HEAT TRANSFER MEDIUM
  (LIQUID METAL)
4 HEAT TRANSFER TUBE/BUNDLE
  FOR CHARGING
5 HEAT TRANSFER TUBE BUNDLE
  FOR DISCHARGING
6 HEAT TRANSFER MEDIUM
7 HEAT INSULATION
8 PRESSURE REDUCING TANK
9 CONNECTION PIPE
10 GAS
11 SUPPLY FOR CHARGING HEAT

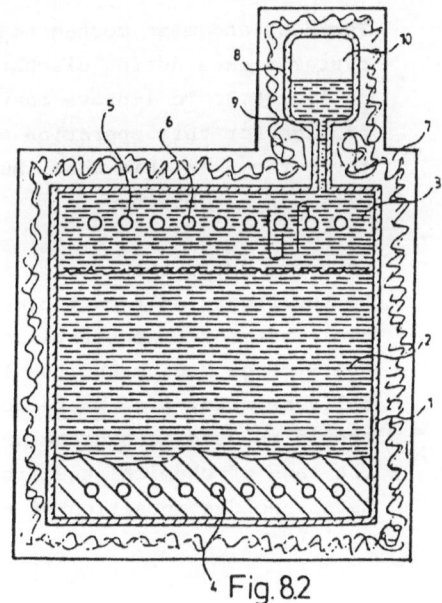

Fig. 8.2

- 253 -

If the temperature difference between the pipes and the
surface of the salt should be small, this is only possible with media showing excellent heat transfer properties. They can be used as layer around and below the
pipes. In particular, liquid metals, e.g. sodium, show
such favourable properties.

Using liquid metals as heat transport fluid, a further
possibility to force a circulation flow within the
transport fluid can be provided by means of liqid metal
pumps, especially of the electromagnetic type. This type
of pump is operating without any mechanical components
in movement (e.g. rotation). Therefore, several problems
normally occuring with active systems, can be limited to
a minimum.

Electromagnetic pumps have to be applied, if natural
convection cannot provide an acceptable heat transfer
flux.

The heat and mass mechanisms for a natural convection
system fluxes during discharging of the system are shown
in fig. 8.3. To improve charging of the system one can
provide for this operation a second heat exchanger at
the bottom. In this case, better heat exchange properties can be obtained.

In every cycle the storage medium is totally mixed, due
to the frozen salt particles settling at the tank bottom. Therefore, no stratification seeems to be possible.
The storage medium and the liquid metal should be chemically related, i.e. preferably a salt of the metal
should be used as storage medium. In this case, due to
the laws of chemical equilibrium and the surplus of
metal, no decomposition of the salt is to be expected.

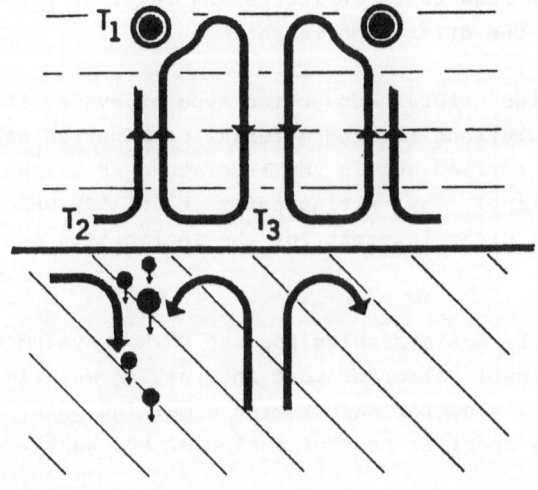

Fig. 8.3: heat and mass transport mechanism at
        the liquid metal/liquid salt surface
        during discharging

The problem of dilatation during the melting process can easily be solved by directing the supply pipe containing the incoming hot fluid or gas vertically through the liquid layer and the salt to the bottom. By this way the melting process starts at the boundary between liquid metal and salt and proceeds along the pipe down to the heat exchanger. There is always space for dilatation, which results in an increasing level of the liquid inside the dilatation vessel.

For the natural convection type of system thermodynamic calculations for the material combination Na/NaJ have been carried out in the literature /8.1/. NaJ shows a relatively low melting point at 661 °C, which might not be of great interest for the application considered here.

But the basic results for the Na/NaJ system seem to be applicable also for more interesting material combinations, e.g. for Na/NaF with a melting point of 1285 K and a specific heat of fusion of 800 kJ/kg.

NaJ is 3 - 4 times heavier than Na, so that no mixture of storage medium and heat exchange medium is to be expected. The main questions of the present storage system are:

- Is the natural convection in the Na-circulation high enough for a sufficient heat transfer rate?
- Can the solid NaJ-particles settle unhindered at high heat output rates?

A rough calculation of the storage system has shown that the natural convection of the heat exchange medium is mainly laminar.

According to the height of the tube-bundle chosen, the heat transfer coefficients were found to be in the order of magnitude of 2 - 4 kW/m^2K.

The temperature differences in the Na-flow are very small compared with the differences between tube-bundle and storage medium. For rough calculations of the heat input and output into and out of the Na-circuit the temperature difference in this circuit can be neglected.

According to the heigth of the tube-bundle chosen a heat input/output rate of about 45 - 65 kW/m^2 can be realized at temperature differences of about 60 K between tube-bundle and storage medium. If the temperature difference decreases to e.g. 15 K, than merely about 12 kW/m^2 can be realized.

For the problem of settling of the solid NaJ-particles a simple calculation model has been set up:

The liquid NaJ will become solid at the NaJ-surface in form of small balls, which sink with constant velocity according to the balance of bouyancy and friction resistance.

In the range of about 50 kW/m^2 which can certainly be realized in a natural convection liquid metal system a ball radius of about $1.5 \cdot 10^{-5}$ m has been found. This radius is small enough that caking at the Na/NaJ-boundary is not to be expected.

8.3   Present state of the art of latent heat storage systems
       at high temperatures

a) State of the art of latent heat storage development

   Studying the numerous patent applications and investigations about latent heat storage systems with
   their different proposals and methods for the enhancement of the storage density, as well as of the heat
   transfer flux from the storage medium to the transport medium, it seems that most of the essential problems, which were identified at the LHS-development
   with temperatures below 1300 K, are solved or resolv-
   able.

   On the other hand, the present state of the art is
   obviously the result of mainly theoretical investigations, without being confirmed by practical application. Practical experience was gained only on the laboratory level.

   Up to now no latent heat storage system for high temperature application (above 400 K) is in practical
   (commercial) operation, although several practical
   investigations and tests are performed in different
   institutes (e.g. for application in private heating
   and solar power plants as short-term-storage).

   The only project beyond the laboratory scale is the
   SOLCHEM PLANT at NAVAL RESEARCH LABORATORIES,
   Washington with:

                2 MWh capacity
                tank ϕ 3,2 m; height 4 m
                storage medium: $MgCl_3$-NaCL-KCL; 27 t

Research and development up to date include:

° theoretical investigations and generation of
  computation programs for adaption of the LHS-system
  on the thermal energy process
° research on melting behaviour of salts and eutectics
° design studies
° material investigations and tests.

In the relevant literature only little practical experience is described, but basic material data are
available. Material data on salt mixtures are incom-
plete.

Essential problems, i.e. the phase separation connected with a decrease of the storage capacity are a result of the "incongruent melting behaviour" of several salt hydrates, used as storage medium at the low
temperature region. They occur only at low temperature LHS systems. Phase separation is not to be expected at high temperature storage systems, because
pure salt melts without crystal water have to be used
as storage medium.

Main problems, which were identified at low temperature storage as well as at high temperature storage
are listed below:

- Stratification, that is the slow segregation (sedimentation tendency) of the seeding crystals. However, the homogenous distribution of the seeding
  crystals is an important condition for the uniform
  discharge of the storage system, without locally
  subcooling and pulsed crystallization.

- Decrease of heat flux from storage medium to external transport fluids during discharge, due to the solidification of storage medium at the "cold" surface of discharging heat exchangers.

- Volume expansion by melting of the storage medium does not cause great problems, if one can avoid an increase of pressure inside the storage tank by means of an appropriate pressure compensation (surge tank).
  However, damage on heat exchanger tube bundles due to the periodic changes of the storage medium volume at charging/discharging have occured.

- Slow segregation of eutectics due to the different specific densities of their component materials after some cycles.

- Chemical instabilitites of the storage medium after a certain number of cycles due to the fact that operating temperature is near to the thermochemical equilibrium of the chemical reactions.

- Finding of an appropriate storage medium and transport medium with regard to the numerous requirements which are to be fulfilled.

The investigation of the internal heat storage and heat release during charging/discharging resulted in the concept of the "direct contact heat exchange (DCHX) in LT-LHS systems. This concept is based on

the use of a heat transport fluid, which is in direct
contact with the storage medium inside the tank and
which is cooled or heated by heat exchangers.

The DCHX, also to be used in high temperature storage
systems, was investigated by R. Tamme, DFVLR /8.2/
and S. Furbo /8.3/. In the related experiments oil
was used as heat transport fluid.

At this development stage the problem of chemical instability of the storage medium was not yet solved,
because the oil is not chemically related with the
storage medium (salt hydrates).

There might be a solution of these problems, if liquid metal is (e.g. sodium) is used as heat transport
fluid, which is also one of the chemical components
of the storage salt (e.g. NaCl; NaJ or NaF).

General advantages of high temperature-LHS using liquid metal as transport fluid and a high melting salt
of this metal as storage medium as compared to low
temperature LHS are

- much higher specific heat of fusion of the storage
  medium (up to one order of magnitude)
- liquid metals used as heat transport medium show a
  much higher heat conductivity than oil or water;
  therefore, higher heat fluxes are possible.
- mass transport by electro-magnetic pumps is possible, also within the storage tank. In this case,
  the latent heat medium can be confined within the
  tank.

b) Experience and know-how on the heat transfer at high temperatures by means of liquid metals

There is a well developed technology of liquid metals, in particular of sodium.

During the fast breeder development, components for sodium storage, heating/cooling and transport (feeding by means of liquid metal pumps) at temperatures up to 900 K have been developed for large scale commercial applications. Components for solar applications have been developed for and have been proven in the solar thermal power plant at Almeria/Spain.

Therefore, up to a temperature of 900 K, proved components and systems are available. But relatively little experience is available on pressurized systems handling the sodium at temperatures above 1150 K (sodium boiling point at 1 bar). This region might be of interest for very high temperature solar systems.

8.4   Advantages and problems to be expected

Advantages of the proposed design (as described in chapter 8.1.2) as compared to former low temperature LHS-systems are as follows:

a) A liquid metal (e.g. Na with melting point at 368 K; boiling point at 1150 K) serves as heat transport fluid within the storage tank. Due to its very high heat conductivity, this fluid could provide a high heat flux from the storage medium to the heat exchanger surface inside the storage tank.

b) Due to the separation of internal and external transport fluids by the heat exchanger walls, no entrainment of storage medium into external components is possible.

c) Although the storage tank can be designed pressureless, the entire storage system is able to provide the pressure transformation function, which is necessary for a wide field of applications in high temperature energy process systems. The pressure transformation is possible due to the separate heat exchangers for charging and discharging which can be operated with different external transport media at different pressure levels.

d) A halogenide of the liquid metal could serve as storage medium (i.e. the salt being a chemical composit of this metal). Due to this fact, no chemical instability of the salt is possible, as long as one of the components (the metal) is present in excess.

e) Damaging of the charging heat exchanger tubes at the tank bottom due to the volume changes of the melting/crystallizing storage medium can be avoided by an appropriate installation of the tubes.

f) For providing a circulation of the internal heat transport fluid, no active systems, such as circulating pumps, stirring by paddels etc., are necessary, because natural convection between salt surface and tube bundles might be high enough to provide a sufficient heat release rate at discharging. In this case, no further problems with active systems will occur.

g) In addition, the use of liquid metal as transport
medium allows the utilization of electro-magnetic
pumps. In this case, the heat transfer rates can be
enhanced.

In spite of the described advantages of the proposed
high-temperature-LHS system, great uncertainties and
lack of experience concerning the physical processes,
which take place inside the storage tank during charg-
ing/discharging, are still remaining. They need to be
investigated:

a) Heat flux achievable during the discharging process.
   - Is the natural convection within the liquid metal
     sufficient for providing heat fluxes, which are
     high enough for technical and commercial applica-
     tion?
   - Enhancement of the heat transfer at the boundary
     between salt and metal by means of additional ac-
     tive systems necessary? (liquid metal pumps?)

b) Physical phenomena concerning the crystallization and
   solidification of storage material:
   - Formation of a solid layer of crystallized storage
     material   at the salt/metal boundary

c) Nucleation phenomena and seeding
   - Possibility of subcooling and pulsed crystalliza-
     tion due to stratification of seeding crystals also
     in high temperature mediums?
   - Segregation of seeding crystals also in high melt-
     ing salts?

d) Is natural convection of the storage medium and movement due to sinking salt crystals sufficient to:
   - achieve a homogenous storage material?
   - solidificate the entire storage medium?
   - avoid stratification?

e) Problems of the tightness of the heat exchanger tube bundles regarding:
   - different transport media
   - different pressure levels
   - high temperatures (1300 K)

f) Compatibility of the salt with the structure material of tank walls and heat exchangers regarding
   - corrosion under material stress
   - high temperatures.

g) Mechanical behaviour
   - Possibilities to avoid damaging of internal structures due to volume changes while melting/solidification of the salt by means of appropriate installation arrangement of these structures
   - Investigation of the mechanical/physical properties of the crystallized salt (i.e. consistency between soft crystal sludge or solid monolithe.

8.5   Proposal for a development program

a) General Approach

In the subsequent development steps, one should solve
the above listed remaining problems and uncertainties.

Up to now basic development work has mainly been executed in the low temperature region of LHS-systems.
The summarized experience gained from this work
should now be used to design a high temperature (HT)
storage system, which can be built and tested within
a real thermal energy process.

Therefore, we propose the storage concept as described in chapter 8.1.2. at first on a laboratory
scale and in a second stage in full size as a prototype plant.

Limiting the efforts to fundamental metallurgical and
chemical research, we propose to use those salts and
salt mixtures which are known already today as a
basis for the design and erection of the test storage
plant. This plant - although on a laboratory scale -
should have its working temperature at the same level
as the full size pilot plant (about 1150 K).

On a small scale the following properties and mechanisms can be tested:

- Compatibility of different structure materials with the appropriate storage and heat transport media, also in long-term tests
- corrosion behaviour of structure material under stress
- tightness of the heat exchanger walls, providing different heat transport media and pressure levels at high temperatures
- variations of operation parameters such as
  ° working temperature (change of the salt melt)
  ° pressure level in storage tank and heat exchanger
  ° charging/discharging level and rate
  ° arrangement of the heat exchangers
- heat transfer rate from salt melt to the liquid metal
- mixing (stirring) of the storage medium due to sinking salt crystals
- volume expansion of the storage medium during charging
- natural convection within the liquid metal
- crystallization and solidification of storage medium during discharging
- mixing effects between salt and metal, i.e. entrainment of storage material, especially if a liquid metal pump is used.

b) Proposed development steps

Systems design work to achieve an optimum integration
of the storage system into an energy process, including the following steps:

- generation of computation programs for the design
  of the LHS-systems itself
- generation of computation programs, which allow the
  simulation of an entire energy system consisting of
  storage, energy source and energy consumers
- definition of requirements to be fulfilled by the
  storage systems applied
- storage design considering the different operation
  modes and technical and commercial aspects
- design and construction work for the test plant and
  prototype plant

Metallurgical and chemical research

We propose a testing program in order to gain experience on structure and storage materials for the development of a LHS-system at a temperature level up
to 1150 K.

Because the practical proving of a storage system
should be the main objective of the development program, the above mentioned fundamental research should
be limited to the following two aspects:

- investigation of the compatibility of different
  storage media with different structure materials
- investigation of the corrosion processes of the
  structure material under material stress and in direct contact to the storage medium.

<u>Design and construction studies</u> as well as erection
of a storage test plant on a laboratory scale, based
on known systems of metals and salt storage media and
on the proposed storage concept, as described in
chapter 8.1.2. The test plant should be integrated
into a closed energy system including an energy
source, storage and sink.

<u>Tests and measurements</u> using the test plant to solve
outstanding problems and to verify theoretical results.

<u>Study and definition of a testing program for storage
systems up to 1200 °C.</u>

As a preliminary work for the development of a LHS-
system at the temperature range between 1150 K and
1500 K the suitability of pertinent storage media and
structure materials should be investigated by a review of pertinent literature.

The same aspects, which are to be investigated within
the test program at the 1150 K temperature level
should serve as selection criteria. For this reason,
the testing plant provided for the 1150 K program
should also be designed as to be appropriate at the
1500 K level.

Basic research should be carried out on pressurized
sodium systems, and also on the possibility of using
copper for very high temperature applications.

c) <u>Design, erection and testing of a prototype storage system</u>

Within the frame of the proposed development steps, design and construction of a prototype storage plant should be performed. These steps also include investigations of the proper integration of such a storage system into an energy process, as well as of operation modes and LHS testing program.

Generally, the testing of a storage system would be advantageous in connection with a complete (integrating) test plant. For example, due to the vicinity to the DFVLR in Porz, the KVK plant, a component test cycle for high temperature gas systems existing at INTERATOM, might be suitable, as well as some other test loops for sodium.

If the test results of the prototype storage plant will be encouraging a full size system, based upon the same technology, should be erected and operated under real technical conditions.

We estimate the necessary effort for the program decribed above on

a) Development of simulation programs
   and respective calculations
   2 scientists, 4 years   80 man-months

b) Material test program
   2 scientists, 3 years   60 man-months
   necessary laboratory equipment   - 300.000,- DM -

c) Design, construction and testing
   of a loboratory scale
   prototype plant   160 man-months
   hardware and operation costs   - 1 Mio DM -

d) Study on advanced systems
   with 1500 K
   2 scientists, 2 years   40 man-months

8.6   First evaluation of development chances

If temperatures below 1200 K are considered, we expect a very good chance for the technical feasibility of the system proposed.

Even for higher temperatures and pressurized sodium good chances for the realization are to be expected. But then external heating of the liquid metal within a respective receiver or heat exchanger is necessary, and problems of storage material transport into the external components have to be solved.

Also copper as liquid metal could be feasible for the storage system itself. But before studying such systems it has to be clarified, whether it is reasonable to store energy on a respective high temperature level.

9. Specific material problems with high temperature solar systems [1]

9.1 Fundamental aspects

When designing solar energy systems, the loading features, which are typical for this source of energy must be used as a basis:

- Clouding cause a large number of thermal-shock transients at the heat-exchanger surfaces (receiver) directly exposed to sunlight.

- Daily start-up and shutdown of the systems cause a build-up of residual stresses and acceleration of creep processes in the components of the pressure retaining boundary.

- The high temperature of the medium results in a greater tendency to creep in the component materials.

As a result of the numerous thermal transients, fatigue occurs so quickly in the component surfaces subjected to sunlight that making provisions in the design for incipient cracking is not feasible.

Where it is accepted that minor incipient cracking will occur on the surface as a result of the thermal-shock loading, the design of the components must feature correspondly low primary stress to counteract crack growth resulting from creeping (high temperature, thermal material expansion). Based for such a design pricipals are shown in fig. 9.1.

---

1) This chapter has been written by Dr. Jansky, BTB

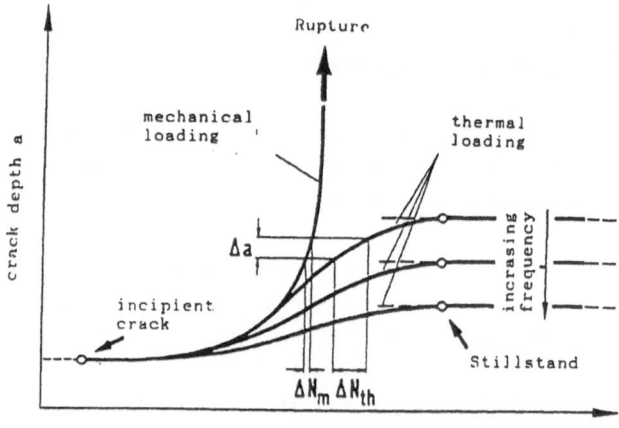

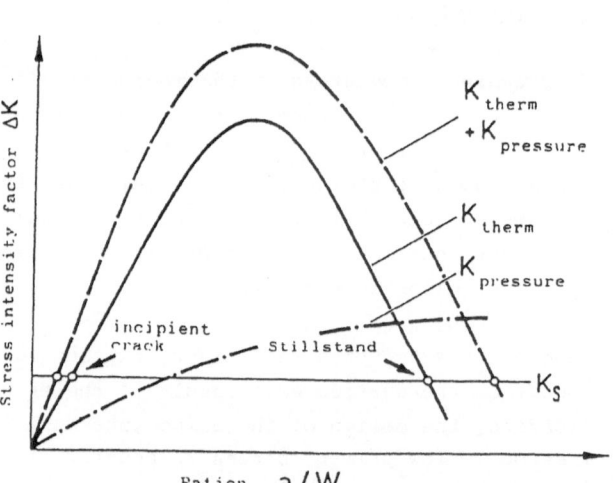

Fig. 9.1: comparison for thermal and mechanical
initiated crack extension under cycle
loading

Notching as a result of geometry and fabrication should be avoided because the dissipation of local increases in stress promotes crack growth.

In addition, the high global forces and torques in the piping and individual components of the pressure retaining boundary, which result from restraints, should be limited so that the possibility of ratcheting deformations can be ruled out.

The pressure transmission medium inside the pressure retaining components must be selected so as to ensure compatibility with the material and all service conditions, i.e. selective corrosion attack of the surface is not possible.

The use of forged components or components which have been put through the pilger mill are recommended for pressure retaining systems. Welds should be optimized so that the moulton weld metal and heat-affected zone of the weld have the same tendency to creep as the base metal. Where the welding process is not optimized, the plastic creep deformations are concentrated in the higher strength zones (usually weld metal) and then these fail prematurely.

Where welding is unavoidable, an effort should always be made to locate the welds in places, which are not subjected to increased local stress/strains.

The increased creep tendency, in other words plastic deformations, resulting from higher temperatures also rule out joints between metals with different coefficients of thermal expansion. Where these cannot be avoided, transition zones must be inserted between both material components.

Change in operating status	Loading created	Reaction of material	Effect	Possible effect on availability
Cloud	* Thermal shock * Change in system temperature and system pressure (relatively slight because of storage action of system)	* Surface fatigue * Build-up of internal stresses (slight) * Fatigue through wall thickness (slight)	Not crack damage Crack growth Crack growth	-- -- --
Start-up and Shutdown	* Change in system temperature and system pressure * Change in restraints	* Creep processes after start-up Reduction of internal stresses * Build-up of internal stresses (after shutdown) * Component wall fatigue * Ratcheting in overloaded sections	Crack growth Overloading of crack tips Crack growth Considerable deformations, crack formation and growth	Leak Slowing up of crack propagation Leak, fracture Leak, fracture
Operation	* High operating temperatures	* Material creep (growth of incipient cracks)	Crack growth	Leak, fracture
At rest during night	* Cold/colder	* Corrosion as a result of high internal stress conditions	Blunting in crack tips	Slowing up of crack propagation

Fig. 9.2: stressing of pressure retaining components in solar energy systems

Similar precautionary measures should also be incorporated for places, where there are seals, e.g. irradiated windows.

9.2   Effect of transients on the pressure retaining boundary

A table in fig. 9.2 summarizes the effects of individual operating conditions. The effect of individual loadings are listed here up to the effect on the pressure retaining boundary.

The design, which always makes provision for fatigue cracking, is based on the fracture mechanics survey of component stress and on the detailed operational inspection of the components.

The fact that the fracture toughness of the material decreases with the duration of loading is reflected in the low level of primary stress.

The leak-before-break system design is desirable. This is to be ensured by numerous in-service inspections.

9.3   Description of development goals

The development goals can be described comprehensively by listing the main fields of problems:

- Selection of materials; use of medium up to 2000 K; material property values (including fracture mechanics values).

- Selection of welding processes and welding parameters (same creep tendency of the weld components).

- Limitation of primary stresses (the goal is 0.5 % expansion after 100.000 hours).

- Limitation of restraints; freely expanding systems (the target is 1 % yield strength after 100.000 hours of operation).

- Fabrication and design adapted to component stress (Finite Element design).

- Inspection of components during test operation (develop measurement technique) and flow of information back to components design.

- Solution of the problems of system joints between metals with different coefficients of expansion (ratcheting deformation).

- Acceptance of incipient surface cracks (with imitation of global deformations) requires the application of fracture mechanics when designing the system components.

- Leak detection under operating conditions for early identification of defects.

- Laying of piping in such a way that stratification of the coolant on the side of the medium cannot occur (be careful with orifaces - feed-in pipes - routing of piping).

- Establishment of fracture mechanics characteristics of materials for the whole service life: accelerated tests.

9.4   Elements of a research and development program

Deducted from the basic aspects and the listing of problems above, we suggest research and development with the
following structure:

9.4.1   Listing of design criteria

One group of materials and design engineer should com-
pile a list of principles for the design of parts of
solar energy systems. For thi⁻, previous operational ex-
perience/breakdowns world-wide should be collected and
evaluated.

9.4.2   Selection and, if necessary, development of materials
for the pressure retaining boundary

The survey should include:

- Compilation of the between 1000 and 2000 K materials
  used to date and an evaluation of their performance
  under operational conditions in solar energy systems
  (creep, corrosion, reduction in fracture toughness);

- Specification of the requirements of the materials in
  accordance with process-related and design data.

- Development of new materials; inspection in laboratories and in use (a modular system for the manufacture
  of the pilot plant is suitable for use).

- Optimizing welding processes and, if necessary, development of new weld components, aiming at the same
  creep properties when subjected to design stress and
  temperature.

9.4.3 Component design and fabrication

The survey should provide:

- Component dimensioning to accomodate stressing, ensuring constant creep deformation, which can be monitored.

  The finite element method is applied here: with actual creep data of the materials (prior simulation of components behaviour).

- Fracture mechanics material characteristics at high temperatures

- Specification of fabrication processes so that the design criteria are complied with;

  Detection of possible fluctuations in fabrication quality as to their effect on the performance of components

- The design of the shape of joints and seals between different material combinations.

9.4.4 Components inspection in operation

A measurement technique for the precise detection of temperatures and deformation/expansion during operation shall be developed.

The aim of the instrumentation is continual monitoring of component use and thus prediction as to where the next in-service inspection shall be performed. This involves centralised data acquisition with a rapid transient recorder.

The same device should also be used for valve monitoring/inspection.

9.4.5   Non-destructive in-service examinations during phases of operation and at stillstand

The following methods shall be applied:

- Leak detection (moisture or noise principle)

- Ultrasonic examinations (dependent on material) - even under operational conditions

- X-ray examinations

- surface examination (metallographic, visual)

In addition, the following procedures are recommended:

- Formation of a group of experts who will offer their opinions on this concept and, if necessary, incorporate additions. Also, this group is to follow up work such as awarding and monitoring of sub-contracts to

  ° Materials, laboratoris or institutes
  ° Specialised welding organisations
  ° Finite element analyses groups
  ° Groups/companies specialising in measurement techniques who cover high-temperature work
  ° Corrosion laboratory (materials/medium connection)
  ° Design offices (CAD), which transfer computer compatible data to the fabrication organisation
  ° Fabricaton workshops which can prove they use computer-aided manufacturing methods.

An overview of recommended activites is summarised in
fig. 9.3.

The project as a whole can be performed in two phases:

- In the first phase, a system will be constructed in
  the materials known to date but using new design
  guidelines; modular construction.

- In the second phase, old parts will be exchanged for
  components made of new materials; and inspection will
  take place under operating conditions, with on-going
  monitoring and supplementation of the design data.

It can be seen clearly from the proposal made, that extensive use of high-temperature solar energy systems is
only possible with careful selection/application of materials and well-considered design/fabrication of the
individual components.

- Selection of design criteria

- Selection and development of
  materials/welded joints

- Component design and fabrication
  suitable for stress criteria

- Measurement techniques for in-
  service component inspection

- Non-destructive examination/
  inspection of components

- Compilaton of examination procedures for fabrication, installation and operation

Fig. 9.3: survey of proposals for implementation
methods/R+D-activities

10.   Market introduction strategies

10.1   Basic aspects

Great importance attaches to the commercial potential of
those solar systems which can be realized economically to
satisfy the long-term power demand of nations which are
situated along the so-called solar energy axis.

A comprehensive analysis of the power market will be conducted taking into consideration

- the actual evolution in national demand up to 2010 for
  most of the countries concerned

- the existing conventional power generators and the future intended power expansion plan of the nations investigated

- long-term price scenarios for fossil fuels such as coal
  and residual oil fuel for the supply of conventional
  power generators

- candidate solar radiation power generators to compete
  with conventional generators

- firm energy of solar radiation of energy producer countries.

Depending on the technical parameters, storage capability
of energy and cost characteristics of the solar power generator type, the potential, which can be realized economically, will be different for each solar technology.

Therefore, a potential study should be carried out, to be used for screening the solar technologies and to identify and promote those technologies, which are of economic interest. It will show, whether or not solar technology can compete with conventional generators and what are the conditions and penalities, which have to be considered in planning a marketing strategy.

10.2   Present and future energy situation in developing countries

A major factor influencing the economic growth of an independent region is the availability of a sufficient supply of electrical energy at lowest total cost. If sufficient indigenous resources are not available primary energy has to be bought on the world energy market. In past years this market has offered energy supplies at relatively favourable prices. However, radical changes are now expected on the world market, characterized by a growing shortage and increasing prices. The international oil market will start this trend most likely within this decade and subsequently other primary energy sources will take the same path. Thus, all countries have to endeavour to become independent of imported primary energy and should try to increase their indigenous resources.

Electricity demand will grow disproportionately over the coming decades.

It is expected that by 2020, the developing countries will account for some 40 % of the world's energy demand. This means that they will triple their present share.

The developing countries must make very intensive efforts
to use their own renewable sources of energy in order to
satisfy their growing needs. Fig. 10.1 shows a map, which
identifies the nations along the solar energy axis. But
solar energy is not only a resource of those countries,
which lay directly inside of the solar energy axis. Fig.
10.1 shows also, that areas with high populaton density
are arranged in the neighbourhood of the solar energy
axis. This gives the chance of exporting solar generated
electricity to consumer countries.

10.3   Research proposal

For finding out the economic potential, we suggest the
application of already existing computer models within a
basic study. Based on Motor Columbus' experiences gained
during planning work for the electricity supply in Africa
and the States of Malaysia a methodology of a long-term
power planning has been worked out.
The power sequence at least cost for the named countries
has already been carried out and has met with wide recog-
nition on the international scene. A flow chart showing
the main activities is shown in fig. 10.2.

Based on the energy demand projections an installation
sequence of power generators can be worked out and optimized towards least cost.

10.3.1 Projections of energy demand

For most of the nations figures for the evolution of
power demand exist. Where this is not the case, one can
prepare a load prediction based on a macroeconomic correlation. This is one of the methods to obtain demand forecast figures.

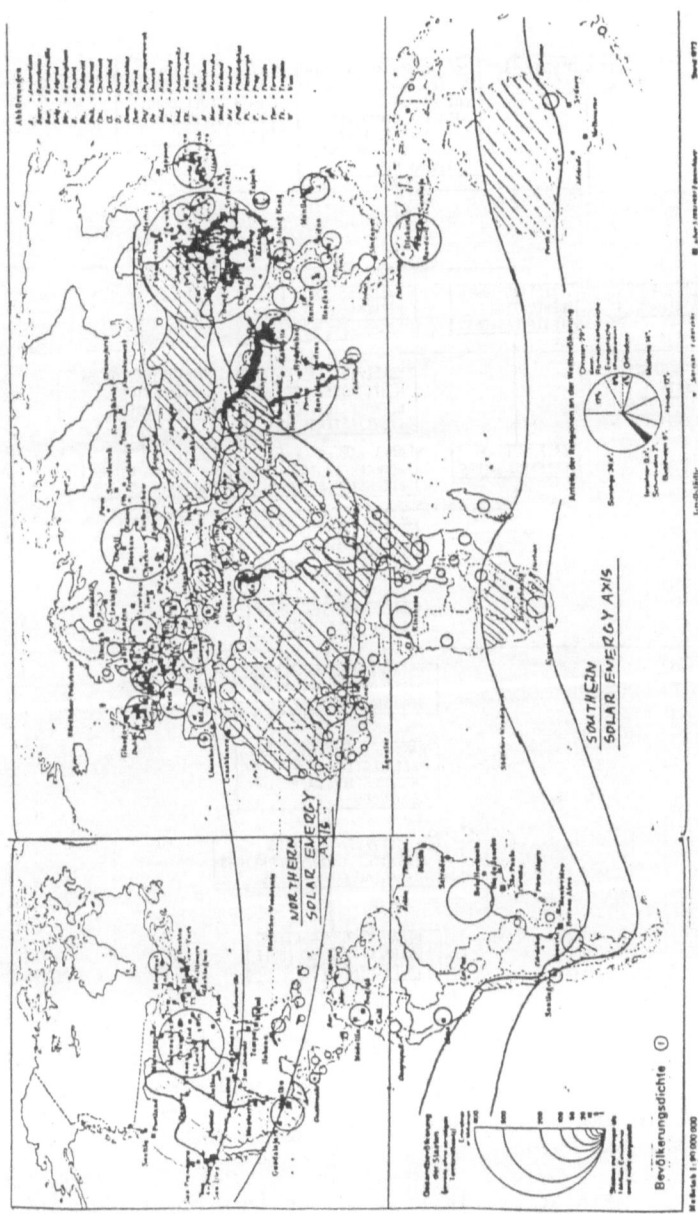

Fig. 10.1: Solar Energy Axis

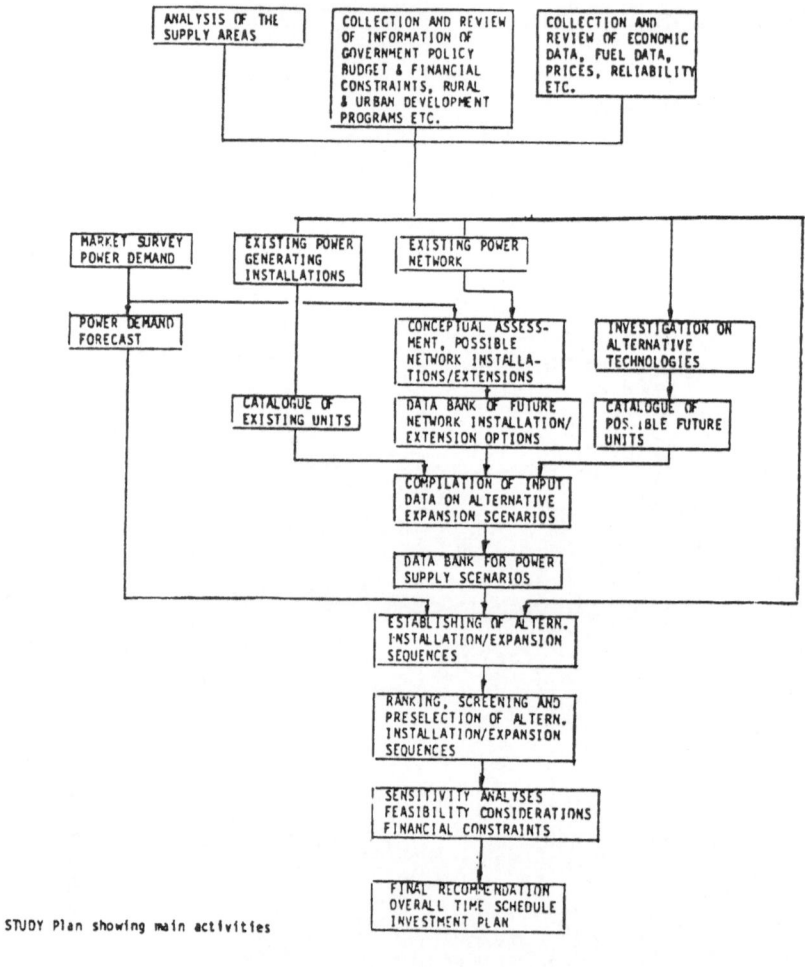

| ANALYSIS OF THE SUPPLY AREAS | COLLECTION AND REVIEW OF INFORMATION OF GOVERNMENT POLICY BUDGET & FINANCIAL CONSTRAINTS, RURAL & URBAN DEVELOPMENT PROGRAMS ETC. | COLLECTION AND REVIEW OF ECONOMIC DATA, FUEL DATA, PRICES, RELIABILITY ETC. |

| MARKET SURVEY POWER DEMAND | EXISTING POWER GENERATING INSTALLATIONS | EXISTING POWER NETWORK |

| POWER DEMAND FORECAST | | CONCEPTUAL ASSESS- MENT, POSSIBLE NETWORK INSTALLA- TIONS/EXTENSIONS | INVESTIGATION ON ALTERNATIVE TECHNOLOGIES |

| CATALOGUE OF EXISTING UNITS | DATA BANK OF FUTURE NETWORK INSTALLATION/ EXTENSION OPTIONS | CATALOGUE OF POSSIBLE FUTURE UNITS |

COMPILATION OF INPUT DATA ON ALTERNATIVE EXPANSION SCENARIOS

DATA BANK FOR POWER SUPPLY SCENARIOS

ESTABLISHING OF ALTERN. INSTALLATION/EXPANSION SEQUENCES

RANKING, SCREENING AND PRESELECTION OF ALTERN. INSTALLATION/EXPANSION SEQUENCES

SENSITIVITY ANALYSES FEASIBILITY CONSIDERATIONS FINANCIAL CONSTRAINTS

FINAL RECOMMENDATION OVERALL TIME SCHEDULE INVESTMENT PLAN

STUDY Plan showing main activities

Fig. 10.2: main activities for a long-term power planning

Methods for predicting future energy demand have been developed, based on its relationship with general economic development. The idea behind these methods is that the use of energy is to a large extent dependent on overall economic growth. The higher the level of production and income in an economy, the higher is usually the use of energy in industry, commerce and households. This can be demonstrated for many countries all over the world. To measure the general economic development of a country, Gross Domestic Product (GDP) figures for constant prices are usually used.

Fig. 10.3 shows a correlation of power demand per capita versus GDP per capita for developing countries. Each point represents a country. During economic development of the country, the point moves upwards parallel to the full line curve.

10.3.2   Data compilation

Based on existing studies for the electricity supply of developing countries carried out by the World Bank, the Arab Fund for Social and Economic Development, the European Commission and KFW, the characteristics of the existing power generation pool and expansion programs can be prepared and compiled for a conventional power generator cataloque.

In the same way the technical parameters and cost characteristics of the solar power technologies can be prepared and compiled for a so-called catalogue of solar technologies.

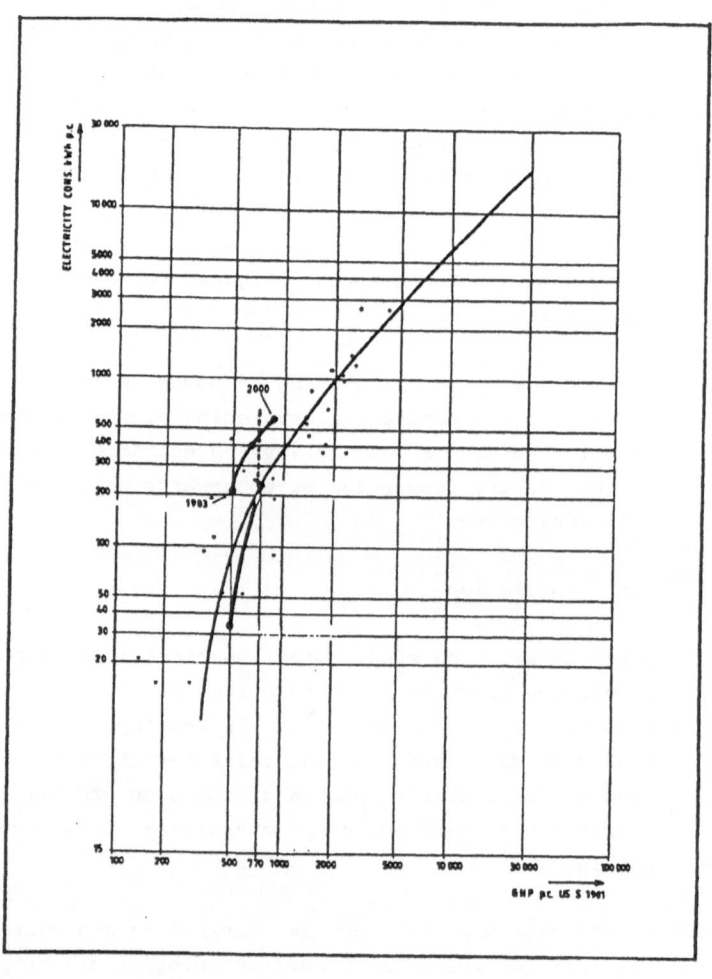

Fig. 10.3: power demand per capacita versus GDP

10.3.3 Future energy supply options

Expansion planning of electric power systems represents dynamic long-term investment problem with annual investments in the order of millions of US-Dollar. Electricity authorities are faced with the task of bringing about fundamental alterations in the electricity economy with the long-term object of making available an adequate supply of low-cost and dependable electricity. In the face of limited financial and energy resources, therefore, the long-term planning of the electric power syste has national importance.

Due to the problem of size and complexity in power systems expansion planning, a mathematical model of a simulation type should be used. Starting with the present configuration of national power system, the procedure is directed towards optimized decisions as regards the commissioning date, as well as type and size of future system additions, i.e. solar power plants and conventional generators. Operated annually to the planning horizon, the model yields a demand-covering system expansion plan, which is evaluated for the present worth of all future systems costs.

Basically there are two options, which compete with each other

- energy supply with conventional generators
- energy supply with a mixture of solar technologies and conventional generators

For the second option a variety of suboptions can be investigated, which consider the various solar technologies with respect to peak load covering generator types and base load covering generator types. Sensitivity analyses will complete the investigation.

The first option consists in retaining the existing demand pattern with an overwhelming portion of fossil fuel at the international market price.

The second option, which is much more favourable, is to change the electricity demand pattern by the substitution of fossil fuels by other energy sources, such as solar radiation.

10.3.4   Commercial Potential

Each project can be described by a financial-mathematical cost and benefit flow, this serving as a basis for assessing economic viability. Optimization is the nucleus of the long-term sequence planning. The most economical way of meeting the demand will be found out by calculating a variety of generating possibilities - solar technologies as well as conventional technologies - and proposals for implementation. A list of priorities for the development of the candidate systems is proposed for the entire sector within the solar energy axis.

The result will be a ranking of candidate solar technologies in order of their exploitation potential.

The study could form a basis for the work of those political bodies and national and international financing institutes, which concern themselves with the electricity supply sector.

Obviously the costs of such a study are directly dependent on the number of countries considered. If only a few number of typical developing countries is studied, some of them preferably being situated close to possible consumers outside the solar axis, we estimate the minimum reasonable effort to about 3 man years.

11.   Concluding remarks

The present states of solar thermal systems has been
derived from technical principles and components, which
were developed for other heat sources. The specific quality of solar energy, which is primarily radiation
energy, was not yet exhausted or even investigated to
the possible extent.

We recommend to concentrate within the next research
period on such fundamental questions. And we feel, that
one should try to become more independent of ideas,
principles and component design, which have been developed historically for quite different boundary conditions.

One should also reflect critically the present status of
some important sub-systems of solar thermal plants,
which , at pesent, are understood within the solar community as "usual" or "reasonable". For instance: Is it
reasonable for technical systems to exhaust merely 2 %
or even less of the theoretical limits?

References

/3.1/   F.K. Boese:
        "Zur Entwicklung thermischer Solarenergiesysteme
        mit sehr hohen Prozeßeingangstemperaturen",
        VDI-Verlag 1985, Kap. 5.2.1

/4.1/   A.Tofighi, F. Sibieude, M. Ducarroir, G. Benezech:
        "Décomposition Thermique a l'Air de la Magnétite
        au Foyer d'un Four Solaire",
        Revue Internat. d. Hautes Témpratures et des ré-
        fractaires, 15, (1978), pp. 1-13

/4.2/   T.S. Laslo, P.E. Glaser:
        U.N. Conference New Sources of Energy,
        Rom 1961, III-F United Nations N.Y. 1964, E/3577/
        rev. 1. Doc. ST/ECA/72, p. 5 and p. 16

/4.3/   E. Bilgen, M. Ducarroir, M. Foex, F. Sibieude,
        F. Trombe:
        "Use of Solar Energy for Direct and Two-Step Water
        Decomposition Cycles",
        World Hydrogen Energy Conference, Miami Beach,
        March 1976, pp. 251 - 257

/4.4/   J.J. Ambriz, M. Ducarroir, F. Sibieude:
        "Preparation of Cadmium by Thermal Dissociation of
        Cadmium Oxide Using Solar Energy",
        Int. J. Hydrogen Energy, 6, (1981)

/7.1/   J.E. Funk, R.M. Reinstrom:
        "Energy Requirements in the Production of Hydrogen
        form Water",
        J. a. EC Process Design and Development, Vol. 5,
        No. 3, July 1966, pp. 336 - 342

/7.2/   G. de Beni, C. Marchetti:
        "Wasserstoff - Energieträger der Zukunft",
        Eurospectra 9, (1979), S. 46 - 50

/7.3/   B.H. Abraham, F. Schreiner:
        "General Principles Underlying Chemical Cycles
        which Thermally Decompose Water into the Elements",
        Ind. chem. Fundam. 13, (1974), pp. 305 - 310

/7.4/   K.F. Knoche, H. Cremer:
        "Thermodynamics of Water Decomposition",
        "The Hydrogen Concept", Ispra, 29. Sept. - 03. Okt.
        (1975)

/7.5/   H. Cremer:
        "Thermodynamische Auslegung und Bewertung von
        Kreisprozessen der Eisen/Chlor-Familie zur Wasser-
        stoffproduktion",
        Habilitationsschrift, Aachen, 15.06.76

/7.6/   H. Hofmann:
        "Die Erzeugung von Wasserstoff mit Hilfe thermo-
        chemischer Reaktionen",
        Chem. Ing. Techn. 48 (1976), S. 87 - 91

/7.7/   E.D. Claudt, A.L. Myers:
        "Hydrogen Production from Water by Means of Chemi-
        cal Cycles",
        Ind. Eng. Chem. Process Des. Dev. 15 (1976),
        pp. 100 - 108

/7.8/   C.E. Bamberger, D.M. Richardson:
        "Hydrogen Production from Water by Thermochemical
        Cycles",
        Cryogenics 16 (1976), pp. 197 - 208

/7.9/ K.F. Knoche:
"Stand und Chancen der Wasserstofftechnologie",
VGB Kraftwerkstechnik 58 (1978), S. 90 - 94

/7.10/ W. Frie, Siemens Forschungslabor:
Mehrere private Mitteilungen
1980/1981

/8.1/ F.K. Boese, H. Weitzenkamp:
"A sodium/salt high temperature heat storage
system",
Colloques internationaux du CNRS, Nr. 306 -
Systemes Solaires Thermodynamiques STS 80 - 109,
1980

/8.2/ R. Tamme:
"Einsatz von Natriumacetattrihydrat und von Barium-
hydroxidoktahydrat in dynamischen Latentwärmespei-
chern",
DGS: 5. Internationales Sonnenforum, 1984, Berlin

/8.3/ S. Furbo:
"Heat Storage with an incongruently melting salt
hydrate as storage medium based on the extra water
principle",
C. den Ouden, Thermal Storage of Solar Energy,
1985

Solar Thermal Energy Utilization
- German Studies on Technology and Application -

Index of Authors

Birke, G.:   Process Synthesis of a Gasification Process
                      Modified for High Solar Energy Integration
                      (Lurgi, Frankfurt), Vol. 3, p. 547.

Bitterlich, W.:   Expert Opinion and Co-operation in the Devel-
                      opment Program High Temperature Storage Tank
                      (Uni-Essen GHS), Vol. 2, p. 211.

Boese, F.:   Considerations and Proposals for Future
                      Research and Development of High Temperature
                      Solar Processes (Motor Columbus, Stuttgart),
                      Vol. 1, p. 169.

Bohn, Th.J.:   Expert Opinion and Co-operation in the Devel-
                      opment Program High Temperature Storage Tank,
                      (Uni-Essen GHS), Vol. 2, p. 211.

Erdle, E.:   Utilization of Solar Energy for Hydrogen
                      Production by High Temperature Electrolysis of
                      Steam, (Dornier, Friedrichshafen),
                      Vol. 3, p. 621.

Freudenstein, K.:   Volumetric Ceramic Receiver Cooled by Open Air
                      Flow - Feasibility Study -
                      (Interatom, Bergisch-Gladbach), Vol. 2, p. 1.

Fuhrmann, H.:   Comparative Investigations and Ratings of
                      Different Solar Systems Using Tubular Steam
                      Reformers, (MAN-Technologie GmbH, München),
                      Vol. 3, p. 251.

Groß, J.:   Utilization of Solar Energy for Hydrogen
                      Production by High Temperature Electrolysis of
                      Steam, (Dornier, Friedrichshafen),
                      Vol.3, p. 621.

Grychta, A.:   Literature Survey in the Field of Primary and
                      Secondary Concentrating Solar Energy Systems
                      Concerning the Choice and Manufacturing
                      Process of Suitable Materials
                      (NU-Tech-Neumünster), Vol. 1, p. 97.

Huber, P.E.:   Considerations and   Proposals   for   Future
                    Research and Development of High Temperature
                    Solar Processes, (Motor Columbus, Stuttgart),
                    Vol. 1, p. 169.

Jäger, W.:   A Multistage Steam Reformer Utilizing Solar
                    Heat, (Interatom, Bergisch-Gladbach),
                    Vol. 2, p. 57.

Josfeld, F.J.:   Expert Opinion and Co-operation in the Devel-
                    opment Program High Temperature Storage Tank
                    (Uni-Essen GHS), Vol. 2, p. 211.

Kalfa, H.:   Layout of High Temperature Solid Heat Storages
                    (Didier, Wiesbaden), Vol. 2, p. 111.

Kalt, A.:   Solar   Steam   Reforming   of Methane   (SSRM)
                    Program Proposals, (DFVLR, Köln),
                    Vol. 3, p. 179.

Kappler, H.W.:   Considerations and   Proposals   for   Future
                    Research and Development of High Temperature
                    Solar Processes, (Motor Columbus, Stuttgart),
                    Vol. 1, p. 169.

Karnowsky, B.:   Volumetric Ceramic Receiver Cooled by Open Air
                    Flow - Feasibility Study -
                    (Interatom, Bergisch-Gladbach), Vol. 2, p. 1.

Kaufmann, J.:   Literature Survey in the Field of Primary and
                    Secondary Concentrating Solar Energy Systems
                    Concerning the   Choice   and   Manufacturing
                    Process of Suitable Materials
                    (NU-Tech-Neumünster), Vol. 1, p. 97.

Koepke, P.:   Yearly Yield of Solar CRS-Process Heat and
                    Temperature of Reaction
                    (Universität München), Vol. 1, p. 3.

Lammers, J.:   Considerations and   Proposals   for   Future
                    Research and Development of High Temperature
                    Solar Processes, (Motor Columbus, Stuttgart)
                    Vol. 1, p. 169.

Lensch, G.:   Literature Survey in the Field of Primary and
                    Secondary Concentrating Solar Energy Systems
                    Concerning the   Choice   and   Manufacturing
                    Process of Suitable Materials
                    (NU-Tech-Neumünster), Vol. 1, p. 97.

Leuchs, U.:   Solar Steam Reforming of Methane - Program
                  Proposals, Vol. 3, p. 195; A Multistage Steam
                  Reformer Utilizing Solar Heat, Vol. 2, p. 57,
                  (Interatom, Bergisch-Gladbach,

Lippert, P.:   Literature Survey in the Field of Primary and
                  Secondary Concentrating Solar Energy Systems
                  Concerning the Choice and Manufacturing
                  Process of Suitable Materials
                  (NU-Tech-Neumünster), Vol. 1, p. 97.

Meyringer, V.:   Utilization of Solar Energy for Hydrogen
                  Production by High Temperature Electrolysis of
                  Steam, (Dornier, Friedrichshafen),
                  Vol. 3, p. 621.

Müller, W.D.:   Steam Reforming of Methane Utilizing Solar
                  Heat, Vol. 3, p. 1; Comparative Investigations
                  and Ratings of Different Solar Systems Using
                  Tubular Steam Reformers (Lurgi, Frankfurt),
                  Vol. 3, p. 251.

Quenzel, H.:   Yearly Yield of Solar CRS-Process Heat and
                  Temperature of Reaction, (Universität München)
                  Vol. 1, p. 3.

Reimert, R.:   Process Synthesis of a Gasification Process
                  Modified for High Solar Energy Integration
                  (Lurgi, Frankfurt), Vol. 3, p. 547.

Siebert, W.:   A Multistage Steam Reformer Utilizing Solar
                  Heat, (Interatom, Bergisch-Gladbach),
                  Vol. 2, p. 57.

Sizmann, R.:   Yearly Yield of Solar CRS-Process Heat and
                  Temperature of Reaction, (Universität München)
                  Vol. 1, p. 3.

Streuber, Chr.:   Layout of High Temperature Solid Heat Storages
                  (Didier, Wiesbaden), Vol. 2, p. 111.

Werner, K.:   Expert Opinion and Co-operation in the Devel-
                  opment Program High Temperature Storage Tank
                  (Uni-Essen GHS), Vol. 2, p. 211.